THE NATURAL HISTORY OF THE COTTON TRIBE

The Natural History

OF THE

Cotton Tribe

(Malvaceae, Tribe Gossypieae)

By Paul A. Fryxell

Texas A&M University Press

COLLEGE STATION AND LONDON

Library of Congress Cataloging in Publication Data

Fryxell, Paul A
    The natural history of the cotton tribe.

    Bibliography: p.
    Includes index.
    1. Malvaceae. 2. Cotton. I. Title. II. Title:
Cotton tribe.
QK495.M27F74   583'.17   78-21779
ISBN 0-89096-071-2

Manufactured in the United States of America
FIRST EDITION

*To the late Thomas Kerr, whose intellectual stimulation
had much to do with directing me toward paths
that eventually led to this volume.*

Contents

Acknowledgments   xi

Introduction   xiii

CHAPTER

1. Historical Background   3

2. Systematics   17

3. Comparative Morphology   103

4. The Origin and Spread of the Tribe   130

5. Ecological Limitations on Spread   150

6. The Influence of Man   159

7. Chromosomal Patterns   178

8. Evolution   185

9. Primitive and Advanced Traits   208

10. Overview   219

Bibliography   227

Index   233

Illustrations

Fig.	1.	Agostino Todaro	6
Fig.	2.	Sir George Watt	7
Fig.	3.	Gavriil Semenovich Zaitzev	8
Fig.	4.	Sir Joseph Hutchinson	10
Fig.	5.	James Otis Beasley	11
Fig.	6.	*Cephalohibiscus peekelii*	21
Fig.	7.	*Cienfuegosia hearnii*	27
Fig.	8.	*Cienfuegosia heteroclada*	28
Fig.	9.	*Cienfuegosia heterophylla*	29
Fig.	10.	*Cienfuegosia affinis*	31
Fig.	11.	*Cienfuegosia sulfurea*	33
Fig.	12.	*Cienfuegosia drummondii*	34
Fig.	13.	*Cienfuegosia ulmifolia*	35
Fig.	14.	*Gossypioides kirkii*	38
Fig.	15.	*Gossypioides brevilanatum*	39
Fig.	16.	*Gossypium sturtianum*	49
Fig.	17.	*Gossypium sturtianum* var. *nandewarense*	50
Fig.	18.	*Gossypium robinsonii*	51
Fig.	19.	*Gossypium triphyllum*	54
Fig.	20.	*Gossypium trilobum*	55
Fig.	21.	*Gossypium thurberi*	57
Fig.	22.	*Gossypium aridum*	59
Fig.	23.	*Gossypium lobatum*	60
Fig.	24.	*Gossypium gossypioides*	61
Fig.	25.	*Gossypium raimondii*	63
Fig.	26.	*Gossypium stocksii*	66
Fig.	27.	*Gossypium areysianum*	67
Fig.	28.	*Gossypium longicalyx*	69
Fig.	29.	*Hampea longipes*	76
Fig.	30.	*Hampea nutricia*	77
Fig.	31.	*Kokia drynarioides*	81
Fig.	32.	*Kokia kauaiensis*	82
Fig.	33.	*Thespesia populnea*	87
Fig.	34.	*Thespesia populneoides*	88
Fig.	35.	*Thespesia acutiloba*	90

Fig. 36. *Thespesia cubensis*   91
Fig. 37. *Thespesia mossambicensis*   92
Fig. 38. *Thespesia grandiflora*   94
Fig. 39. *Thespesia multibracteata*   95
Fig. 40. *Thespesia fissicalyx*   96
Fig. 41. *Thespesia robusta*   96
Fig. 42. *Thespesia garckeana*   97
Fig. 43. *Thespesia gummiflua*   98
Fig. 44. Embryos of *Hibiscus* and of genera of the Gossypieae   104
Fig. 45. Cross-section of seed of *Gossypium*   105
Fig. 46. Leaf outlines of members of the Gossypieae   109
Fig. 47. Domatia as found in *Thespesia cubensis* and
          Hampea tomentosa   112
Fig. 48. Buds of members of the Gossypieae   115
Fig. 49. Staminate and pistillate flowers of *Hampea nutricia*   120
Fig. 50. Flowers of *Cienfuegosia hildebrandtii* and *C. affinis*   122
Fig. 51. Geographical distribution of the genera of the Gossypieae   131
Fig. 52. Geographical distribution of *Cienfuegosia yucatanensis*   143
Fig. 53. Diagram of the origin of tetraploidy in *Gossypium*   163
Fig. 54. Geographical distribution of the tetraploid species of
          Gossypium   165
Fig. 55. Diagrammatic branching sequence for the phylogeny of the
          diploid species of *Gossypium*   215

TABLES
Table 1. Effects of Changing Requirements on the Cotton Crop.   174
Table 2. Changing Patterns of Selected Characters through the
          Course of Cotton Domestication.   175
Table 3. Reported Chromosome Numbers in the Tribe Gossypieae.   179
Table 4. Primitive and Advanced Character States of the Genera of
          Gossypieae.   212
Table 5. Similarity Indices Derived from Table 4.   214

Acknowledgments

IT is impossible for me to acknowledge all the debts that I owe different people for material assistance, intellectual stimulation, and other contributions made in connection with this work. Many of the ideas expressed in these pages have had complex origins, and I cannot claim full credit for them. Because of these complexities I can neither trace the origins of these ideas nor allot credit where it may be due. I can only ask forgiveness of those who may have been slighted.

A few special acknowledgments do need to be made, however. I am grateful to Mrs. Adele Lewis for patience and skill in typing the manuscript, to Dr. John Endrizzi for a critical reading of the manuscript and many helpful suggestions, and to Dr. Greta A. Fryxell for support and encouragement during the developmental stages of this project and for many useful discussions, especially on taxonomic concepts.

I appreciate the help of Mr. Nikolai Lemeshev for supplying the portrait of Gavriil S. Zaitzev, and I am grateful to the Hunt Institute for Botanical Documentation for permission to use the portraits of Sir George Watt and Sir Joseph Hutchinson. The portrait of J. O. Beasley is from the archives of the Evans Library of Texas A&M University, to which appreciation is expressed.

I am grateful to a number of organizations and individuals for permission to reprint illustrations as follows: the University of Chicago Press (Fig. 44); the New York Botanical Garden (Figs. 7, 13, 49); the Missouri Botanical Garden (Figs. 7, 50, 52); the Muséum National d'Histoire Naturelle, Paris (Figs. 15, 34, 43); Mr. J. H. Saunders (Figs. 16, 18, 22, 25, 26, 28); Dr. P. Valíček (Figs. 17, 19); the California Botanical Society, *Madroño* (Fig. 23); the Escuela Agricola Panamericana, *Ceiba* (Fig. 29); Dr. José Sarukhán (Fig. 30); Dr. Otto Degener (Figs. 31, 32, 33); the editorial board, *Flora Zambesiaca* (Figs. 35, 37, 42); Dr. R. A. Howard (Figs. 36, 38); and Dr. J. van Borssum Waalkes (Figs. 39, 40, 41). The firm of Bernard Quaritch

Ltd., London, was kind enough to facilitate use of the Redouté drawing (Fig. 9), and the Fogg Art Museum of Harvard University likewise assisted in obtaining usable reproductions of the St.-Hilaire plates (Figs. 10 and 11).

Introduction

Believing that it is always best to study some special group, I have, after deliberation, taken up domestic pigeons.
                                                    Charles Darwin

Dr. Fryxell, you come with mallows aforethought.
                                        paraphrased from S. R. Hill, 1972

In Support of Monographic Work

A question posed in philosophical circles is, Which came first, the chicken or the egg? In the field of plant taxonomy the parallel poser is, Which is more important, floristic studies or monographs? A monograph concerns itself with all the species of a given higher taxon, say a genus or perhaps a family. A flora, by contrast, deals with all the species of a given area, such as a natural geographical area, a political unit, or an ecological habitat. The answer to the question, of course, is that neither is "more important" than the other; rather, they are mutually supporting.

But since taxonomists are human, they sometimes adopt more partisan views, and their answers to the question relate to whether their own activities tend to emphasize the writing of floras or of monographs. I wish to detract in no way from the important work of those, myself occasionally included, who produce floristic studies. But I do wish to say a word, in part as a justification of the present work, about the positive significance of broadly based monographic studies (cf. Stuessy, 1975).

Taxonomy is a two-legged science, being made up of the complementary but distinct activities of classification and identification. Classification is concerned with taking a group of organisms and systematizing it—that is, devising a scheme of classification or a taxonomic concept for the better comprehension of the group and its members. It is the best method yet devised for the storage and retrieval of biological information. The end result is a group of names that identify the units of classification, arranged in a hierarchical fashion.

The group may be defined taxonomically, as in a monograph, or geographically, as in a flora. The two approaches become identical for those botanists capable of dealing with a general system—that is, of devising a classification system for "plants of the world."

The process of identification is concerned with fitting a given plant specimen into an existing system of classification—with identifying and naming an unknown plant. Necessarily, the construction of floras is a process that is more concerned with the identification part of taxonomy than with the classification part; at least this tends to be true in practice. Conversely, monographic work is primarily concerned with classification, even though it must also concern itself with problems of identification. A complete monograph includes a key for the identification of the entities classified. But my point here is that there is a difference in emphasis.

A third leg of taxonomy is nomenclature. Perhaps I should call it the tail, using the analogy of a kangaroo, since it is entirely supportive of the other two. Nomenclature is of no importance for its own sake, but takes on importance because both classification and identification require a stable, precise, and workable system of nomenclature. It may be *of interest* for its own sake, because of the precision and complexity of the nomenclatural system that has evolved and been codified, but its importance is in its supportive role as the necessary aid to communication about plants, especially about the classification and identification of plants.

Why, then, do I wish to emphasize the importance of monographic work? Partly, no doubt, it is simply a matter of personal taste on my part and as such should be simply asserted and accepted, not defended. But it goes deeper than that. To me, a classification of the biological world (or any portion of it) can be meaningful only if it is projected in a phylogenetic context. No subject can be fully understood apart from its history. I recognize, as some taxonomists are at great pains to point out, that (1) a phylogeny can only be inferred, and the inferences are often based on relatively indirect evidence, except in those rare instances where there is a fossil record, and (2) systematics *need not* be phylogenetically based, but may be based on whatever system of criteria may serve a particular purpose. Let me discuss each of these points in turn.

The first argument I find weak because it simply says, "If I can't have all the candy, I don't want any at all!" Of course we can never know with certainty the actual phylogenetic history of the vast majority of plant groups. But we can learn a great deal by inference, and often the resulting knowledge can have an important bearing on broad botanical problems. Moreover, there are rare instances in which verification has been possible, and the projected phylogeny has been proved correct (for example, the projection by Mangelsdorf of the primitive ancestor of maize and its subsequent discovery at Tehuacán), thus giving confidence in the criteria and methodology used in making the projections.

The second argument is certainly correct as far as it goes. Taxonomy is a field of inquiry much broader than its biological application. Classification of objects, of concepts, of events, of persons is an integral part of all human activity—indeed, it may be argued that the ability to classify is a basic attribute of the human condition. As has often been pointed out, a given group of entities, whether it be a bin of machine bolts, a flight of military aircraft, or a collection of poems, may be classified in a variety of different ways, all equally valid, depending upon the purposes for which the classification is made. This is certainly true of plants as of anything else.

The point at issue here, however, is not what *may* be done in classifying plants, but what *is* done and why. Those who argue that classification systems need not (indeed, should not) have a phylogenetic rationale seem, for one thing, to make the tacit assumption that the only or principal purpose of a classification system is identification. Their claim that a classification is to be judged only in terms of its usefulness for this purpose misses the mark. Usefulness is not an objective of the taxonomic enterprise—rather, it is a by-product, albeit a by-product to be sought after earnestly. The principal objective of the taxonomic enterprise is to arrive at a classification that has the property not of usefulness, but of naturalness. Usefulness cannot be a property of a classification; it is a property of the *means of presentation* of a classification (keys, descriptions, illustrations, and the like). If a classification achieves its primary objective, naturalness, then the taxonomist has the opportunity to achieve the by-product as well, but this is a separate task using different talents.

I assert that the purposes of classification are both broader and deeper than identification alone and must be aimed at a comprehension of the plants and at achieving a classification that effectively serves as an information storage system. Such comprehension cannot ignore the history of the plants even though that history may be dimly perceived. Phylogeny must be basic to plant systematics if systematics is to achieve its primary goal of natural classification.

It has been said that you can prove anything you want in phytogeography by selecting your examples. This criticism can be met by basing your conclusions on a total sample rather than a selected sample—that is, by using a monographic approach to plant geography. My defense of monographic work, therefore, is based upon the greater power of such studies to grapple with phylogenetic and phytogeographic problems. It is supported by the greater emphasis on classification than on identification inherent in monographic work as compared to floristic studies. The classificatory emphasis, to me, provides a greater challenge and involves a greater element of creative activity and conceptual innovation. For all of these things to be achieved, a monograph must be broadly based. "The Genus *Gossypium* in the United States" would have little penetration of significant problems and would provide little insight into the genus. Even "The Gossypieae of South America" would provide a misleading picture of the group as a whole. Preferably, the scope should be worldwide and should include as large a taxonomic group as it is possible for the investigator to include. That is the aim of this book.

In Support of Natural History Studies

ALTHOUGH this volume uses a taxonomic presentation as its framework, its purpose is broader than the simple presentation of a classification system for the group. As has been implied, phylogenetic considerations will receive a major emphasis. But the term *natural history* implies even more, and the intention is to approach the topic from such a broadly based orientation.

There is an area of biological investigation, the study of evolution, in which the following three fields of specialization intersect or overlap: (1) the study of internal mechanisms of evolution (genetics); (2) the study of the external interactions of evolving organisms (ecology); and (3) the study of the systematic relations among the products of evolution (taxonomy). It is this area that I have chosen to call "natural history" in the title of this work. It is, in my opinion, extremely difficult to divorce these three aspects of biology from one another. Natural history is a multifaceted but unitary topic.

The Gossypieae are considered to merit this intensive treatment for a variety of reasons. The most obvious reason, perhaps, is the overriding economic importance of the type genus *Gossypium*, which has resulted in a great backlog of knowledge about these cultigens and to a lesser extent about their wild relatives. The very naturalness of the group, at as high a rank as tribe, not only makes the group convenient to study but also promises that significant results may be achieved. Its worldwide distribution and its ancient age emphasize this promise. Added interest results from the inclusion of both successful and relict genera in the tribe as well as successful and relict species (cf. Fryxell, 1962).

This monographic approach to the study of evolutionary processes and patterns is complementary to the more usual eclectic approach that draws examples from diverse but intensively studied organisms (*Drosophila, Gilia, Triticum,* and the like) which often have been intensively studied (as has *Oenothera*) because they are atypical. The monographic approach to evolutionary studies, by dealing with a total sample rather than a selected sample, seeks a balanced view of the evolutionary patterns found in the group studied.

What is a natural history? Constance's phrase "systematic botany an unending synthesis" (Constance, 1964) aptly captures the essence of the methodology of research in systematics and at the same time characterizes the taxonomic aspects of natural history. But natural history is a bit more inclusive than systematics alone. It encompasses evolutionary history, coevolutionary relationships, community ecology, and various related considerations such as dispersal mechanisms, reproductive biology, and niche adaptation. In these areas, research methods are more analytic than synthetic. One recalls Smith's phrase "systematics and appreciation of reality" (Smith, 1969) as an expression of this broader outlook. I prefer to think of natural history in terms of the *perception* of reality, of an attempt to grasp "what is out there" in the real biological world. Both analytic and synthetic approaches are needed.

Descriptive biology has a real and important place in science and human aspirations. It is not a dusty aberration of myopic specialists and eccentric amateurs more attuned to the time of Linnaeus than to the contemporary world, even though many contemporary biologists who are more at home with laboratory equipment than with biological organisms seem to believe that is true. Rather, descriptive biology is a matter of urgent business, made more urgent by the accelerating pace of destruction of the biological world by human population pressures. Descriptive biology has as its goal the knowledge and understanding of the totality of the biological world that is "out there" but of which we are so intimate and interdependent a part (cf. Fosberg, 1972; Evans, 1973). It has a long way to go to accomplish that goal, and much work remains to be done. In getting on with the job, I prefer the "cosmic optimist" viewpoint of Smith (1969) to the pessimistic view of Raven et al. (1971), although this optimism is perhaps better warranted with some groups of organisms than with others.

In any case, this book attempts to place a particular group of organisms, the cotton tribe, in the context of its evolutionary history, its phytogeographic distribution, its ecological range, its interaction with man—in a word, its natural history.

THE NATURAL HISTORY OF THE COTTON TRIBE

1

Historical Background

We do not hold Gossypium *to be so difficult a genus as it is generally represented to be. We in northern Europe can do little towards working it up, but a botanist of average ability residing in some tropical or semitropical country could easily put it to rights.*

Berthold Seemann, 1866
(in review of Parlatore's
monograph)

For a few of the most important crops such as cotton and wheat, a series of monographs has shown that the problem of the origin of cultivated plants is immensely bigger than it seemed to DeCandolle and the botanists of his day. It will take a series of monographers merely to outline the job he was once thought to have finished fairly acceptably.

Edgar Anderson, 1969

THE tribe Gossypieae is a well-defined unit within the family Malvaceae, and yet its recognition as a natural and coherent taxon has been slow in developing. The family Malvaceae is readily divided into two groups: plants with schizocarpous fruits such as *Althaea* and *Pavonia*, and plants with capsular fruits such as *Hibiscus* and *Gossypium*. The latter group was segregated as a distinct tribe by Reichenbach in 1828 and has been consistently recognized by essentially all subsequent workers. The concept that Reichenbach presented, however, combined what are today recognized as two distinct tribes, the Hibisceae and the Gossypieae. It is the distinction between these two that was gradual in its development.

Alefeld (1861) was the first to perceive and delineate the tribe Gossypieae. In describing the tribe, he emphasized two characters: the form of the embryo (which is more complex than in the balance of the Malvaceae) and the presence of distinctive punctae in various parts of the plant but especially in the cotyledons. These punctae are now known as "gossypol glands" and are distinctive in morphology and chemical contents. They are believed to be unique to the tribe.

Alefeld's conception of the tribe was perceptive—and ahead of its time, because his contemporaries and successors did not grasp the concept. Bentham and Hooker, in their great *Genera Plantarum* (1867), divided the tribe Hibisceae into two "series," one centering on *Hibiscus* and the other on *Gossypium*. They did not, however, use Alefeld's characters to distinguish the groups, but rather used the length of the style branches and the shape of the seeds. The former character is a supportive one at best, and the latter is an imperfect reflection of the underlying differences in embryo structure. As a result, Bentham and Hooker underestimated the naturalness of the tribe Gossypieae and did not recognize it in tribal rank. The views of Bentham and Hooker were so influential, however, that they were followed by most subsequent workers.

Dumont (1887) was one exception. He recognized the taxonomic value of the gossypol glands and added distinctive characters of wood anatomy that served to characterize the Gossypieae. Watson had also recognized, but only in passing, the taxonomic value of the gossypol glands occurring in the cotyledons. Ulbrich (1914) and later J. B. Hutchinson (1947a) then presented an implicit conception of a natural group of genera allied to *Gossypium* and distinct from the Hibisceae sens. str. Reeves (1936) added anatomical characters, especially of the seed coat, that further supported the distinctiveness of the tribe Gossypieae.

But in spite of this developing picture and of the reinforcements to Alefeld's conception that appeared in succeedings years, most botanists stayed with Bentham and Hooker's view that the tribe Hibisceae may have two weakly differentiated "series" but is only a single tribe (for example, Schumann and Gürke, 1891–1892; Kearney, 1951; J. Hutchinson, 1967; Edlin, 1935; J. B. Hutchinson, 1947a; Corner, 1976). Such is the long shadow cast by authority. Alefeld was eminent, but Bentham and Hooker were more eminent.

The conception of Alefeld was resurrected, however, and I brought together subsequently published data that bore on the delimitation and characterization of the tribe Gossypieae (Fryxell, 1968a). That synthesis produced a modern treatment of the tribe Gossypieae that is the basis for this book.

The tribe comprises eight genera that are listed and described in chapter 2. The balance of this historical background will discuss each of the genera individually.

Gossypium

Gossypium is a Linnaean genus and was the first of the tribe to be described. Linnaeus originally included three species in the genus and later described two more (Fryxell, 1968*b*). Swartz (1790) was the first to treat these species as a group. Additional species have been described since that time up to the present day, giving rise to a welter of names (675, according to Fryxell, 1976) that reflect the great diversity to be found in this genus, which includes several species that are highly variable as a result of their long history of cultivation.

The genus first received serious taxonomic attention in the mid-nineteenth century when the Italian botanists Parlatore and Todaro (Fig. 1) published the first monographic treatments of *Gossypium* (Parlatore, 1866; Todaro, 1863–1864, 1877). Their attention was directed to *Gossypium* because of the introduction of cotton as an agricultural crop into Italy at about that time. One must qualify this Italian priority, however, with reference to the earlier work published by von Rohr (1791–1793), which was based on his studies of cotton in the Danish West Indies during the 1780's. The first extensive collection of living material of *Gossypium* was established by von Rohr at St. Croix, and his book recorded his observations on these plants and a classification of them based almost entirely on seed and fiber characters, that is, characters of agricultural importance. Although von Rohr's understanding of these cottons was great, his descriptions are too sketchy to have much systematic significance, and he provided only vernacular names. Rafinesque subsequently supplied the missing Linnaean binomials (Fryxell, 1969*a*), but his descriptions were merely borrowed from von Rohr, and his understanding of the plants was minimal at best. Thus, neither singly nor together can the efforts of von Rohr and Rafinesque be said to constitute a taxonomic monograph of *Gossypium*. The nod of priority has to go to Parlatore and Todaro.

Parlatore's treatment was relatively brief. Although it was comprehensive in scope and characterized by thoroughness, it was basically a literature review and history of the genus. Todaro's study of *Gossypium*, on the other hand, was characterized by originality and insight as well as by thoroughness. He assembled a large collection of *Gossypium* from all parts of the world and studied it in living condition in the Royal Botanic Garden at Palermo. His several publications on *Gossypium* (see Fryxell and Smith, 1972) culminated in his 1877

Fig. 1. Agostino Todaro (1818–1892).

monograph, which was a fully matured interpretation of the genus. It was the first major landmark in *Gossypium* taxonomy.

The next significant contribution to our understanding of *Gossypium* systematics was the doctoral thesis of Angelo Aliotta (1903), who extended the Italian monopoly of the study of the genus. Aliotta's work in some respects is simply an updating of the works of Parlatore and Todaro. It is characterized by a clear understanding of the importance of documentation, and Aliotta's observations and interpretations are carefully based on actual herbarium specimens (often the same specimens studied by Todaro), not only on literature sources.

Aliotta's work, however, was quickly overshadowed by the second major landmark in the study of *Gossypium* taxonomy, the monumental

Fig. 2. Sir George Watt (1851–1930). (Courtesy, the Hunt Institute for Botanical Documentation, Pittsburgh, Pennsylvania)

Wild and Cultivated Cotton Plants of the World by Sir George Watt (1907). Watt (Fig. 2) was an English botanist who resided for many years in India, where he acquired a firsthand knowledge of cotton and a continuing interest in the subject. Somewhat earlier (1893) he authored the well-known *Dictionary of the Economic Products of India*. His monograph on cotton is an unusual storehouse of historical and botanical information, much of it not readily available from other sources, including photographs of type specimens and other important illustrations. For this reason Watt's work will continue to be consulted long after his outdated taxonomic conceptions have been relegated to the category of "historical interest only."

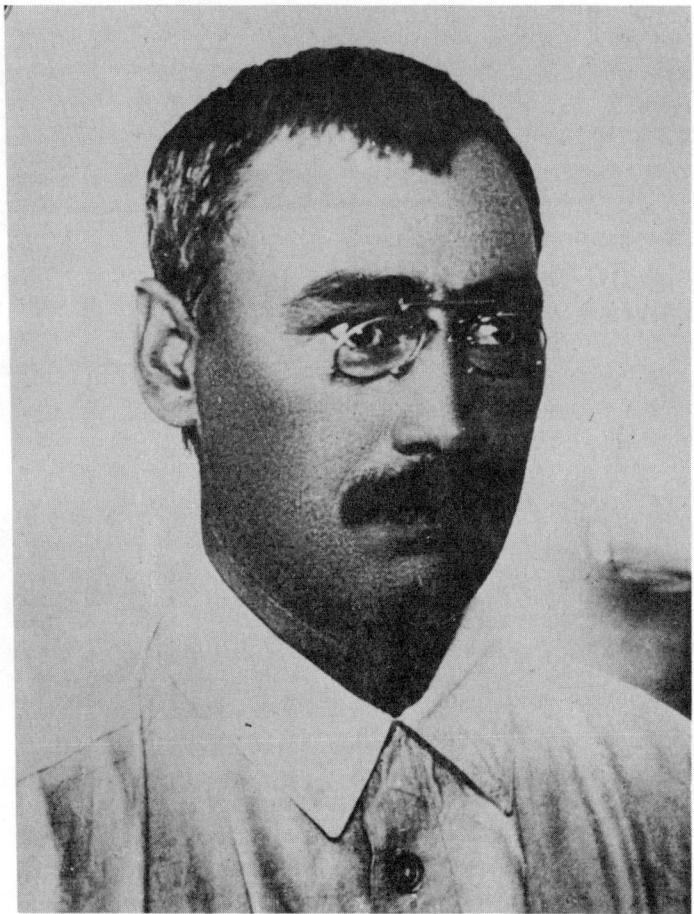

Fig. 3. Gavriil Semenovich Zaitzev (1887–1929).

We now come to the person who, in my opinion, deserves the place of honor in the history of *Gossypium* taxonomy, the Soviet agricultural botanist G. S. Zaitzev (Fig. 3). From a historical perspective, the taxonomic treatments of *Gossypium* may be placed in two groups: those treatments that were based on the important insight of Zaitzev, and treatments by those botanists who preceded Zaitzev or who failed to understand or accept his contribution. The watershed of our understanding of the taxonomy of *Gossypium* is Zaitzev's now classic paper, "A Contribution to the Classification of the Genus *Gossypium* L.," published in 1928.

Zaitzev confined his attention to the lint-bearing (that is, the culti-vated or potentially cultivated) cottons. He took a broad overview of their ecology, phytogeography, and cytology and perceived that they could be divided into two groups, the Old World cottons and the New World cottons. The Old World cottons are diploid ($2n = 26$), and the New World cottons are tetraploid ($2n = 52$). They hybridize only with difficulty, and the hybrids, when produced, are completely sterile. Before Zaitzev's contribution, botanists such as Todaro and Watt had been so preoccupied with the myriad variations shown by the culti-vated cottons and with describing, naming, and classifying them that they failed clearly to perceive this elementary relationship. This is not to be critical of Todaro or Watt, for they worked in the context of their times, but it adds to the luster of Zaitzev that he transcended their thinking.

Next, Zaitzev noted that each of these two major groups could itself be subdivided into two subgroups—the New World cottons into Central American and South American subgroups, and the Old World cottons into African and Indo-Chinese (or Asiatic) subgroups. He thus resolved the manifold variation of the lint-bearing cottons into four major groups, each with morphological and geographic integrity. It was a bold stroke, but subsequent understandings have shown him to be correct in his insight.

Zaitzev's interpretations were in terms of the Vavilovian concepts of centers of origin and of the genetics of homologous series. He had the advantage of these Vavilovian theories and of developing knowl-edge of cytogenetics that his predecessors lacked. He did not make his 1928 paper into a taxonomic treatise, but only a means for presenting his concepts. His opening paragraph stated that he planned in the future to publish "a special monograph of the genus *Gossypium,*" but it was never published. It remained for others to build on his foundation, for he died on January 17, 1929, a tragic loss to botanical science (Vavilov, 1929). He was but forty-one years old.

Zaitzev's four "groups" of cottons were never stated by him to be anything (in a taxonomic sense) other than "groups" of "forms." I pre-sume he was intentionally withholding taxonomic judgment at that time so as not to detract from the presentation of his significant simpli-fying insight or to anticipate his proposed monograph. Almost twenty years later, J. B. Hutchinson (now Sir Joseph Hutchinson, Fig. 4) perceived Zaitzev's genius and followed his thinking to the end of

Fig. 4. Sir Joseph Hutchinson (1902–). (Courtesy, the Hunt Institute for Botanical Documentation, Pittsburgh, Pennsylvania)

reducing each of the four groups to a single species. Hutchinson's treatment (Hutchinson, 1947*b*) has been widely and generally accepted.

Zaitzev never published his promised taxonomic monograph of *Gossypium* because of his unexpected and untimely death. However, his favorite student, F. M. Mauer, published such a monograph in 1954 (critically reviewed by Varuntsyan, 1958) that, like Hutchinson's, was based on the concepts of Zaitzev. In this way, through his student a generation later, Zaitzev's promised monograph came to fruition.

Fig. 5. James Otis Beasley (1909–1943). (Courtesy, Texas A&M University Archives)

The monographs of Hutchinson and of Mauer benefited from new knowledge developed during the decades following Zaitzev's death, especially new understandings in cytogenetics. A key position in this development is occupied by the classic studies of the American botanist J. O. Beasley (Fig. 5), who established the different genome groups within *Gossypium* that are the foundation for modern taxonomic interpretations of the genus and who demonstrated the amphidiploid origin of the New World tetraploid cottons (Beasley, 1940*a*, 1940*b*). Another tragic loss to botanical science occurred when Beasley was killed in World War II military action at the age of thirty-four. (The Beasley Laboratory of cotton cytogenetics at Texas A&M University is named in his honor.) Beasley's contribution to our understanding of the phylogeny of *Gossypium* rivals Zaitzev's insight into the taxonomy of the genus. These two key contributions were brought into focus in the works of Hutchinson and Mauer.

This historical summary of *Gossypium* can be brought to a close by brief references to relatively recent publications. The detailed and elaborate classifications of *Gossypium* presented in a series of publications by Roberty (1942, 1946, 1950) scarcely deserve comment and can be simply dismissed as unsatisfactory and unusable. Other studies (for example, Gammie, 1905; Harland, 1940; Konstantinov, 1939; Wouters, 1948) are too minor to merit comment or are only peripherally concerned with taxonomic questions. It is remarkable that T. H. Kearney, a trained and capable taxonomist who devoted much of his career to studying *Gossypium*, produced no major taxonomic study of this genus.

In 1947 Prokhanov published a "conspectus" of the genus, the historical antecedents of which are of some interest.* Prokhanov claims that "the unforgettable N. I. Vavilov... personally entrusted [to] me, as a botanist, the detailed study of the taxonomy of *Gossypium*. He did not believe that Mauer, an epigone of the well-known Zaitzev,... would do this job in an adequate manner." Zaitzev, however, on his deathbed designated Mauer to be his successor (personal communication from M. G. Zaitzeva, daughter of G. S. Zaitzev), and it is difficult to suppose that Mauer succeeded his mentor without at least the tacit approval of Vavilov, since Mauer devoted a major portion of his career to the study of the living *Gossypium* collection at Tashkent, assembled through the efforts of Vavilov and S. M. Bukasov. When Prokhanov visited Tashkent to study this collection in the early 1930's, relations between him and Mauer were evidently strained. Varuntsyan (1958) comments further on continuing friction between Prokhanov and Mauer. "For various reasons" Prokhanov discontinued studying *Gossypium* until 1944, when he resumed work "preparing a [doctoral] thesis 'The Cottons of the World and Their Wild Relatives' [which] was defended in Leningrad at the Botanical Institute in 1949." Prokhanov's

---

*Parts of this story are derived from a letter Prokhanov wrote to me dated October 11, 1964, shortly before his death. The quotations are from the letter.

conspectus is extracted from his thesis, but except for the conspectus and a paper on the identity of *Gossypium barbadense* L., also extracted from the thesis and published in 1959 (cf. Fryxell, 1968*b*), the thesis remains unpublished. Prokhanov (in 1964) indicated that he planned further publications, in part to refute Mauer, but these did not materialize because his death intervened.

Prokhanov's taxonomic treatment of *Gossypium* breaks completely with the modern views that are based on Zaitzev's conceptions and reverts to the earlier approaches of Todaro and Watt, making a multiplicity of species from among the lint-bearing cottons. Todaro recognized thirty-nine lint-bearing species; Watt recognized nineteen and later (1926, 1927) described an additional ten; Prokhanov recognized forty-one species that were broken up into a plenitude of infraspecific taxa. By contrast, Hutchinson accepted five species and Mauer seven in describing these plants. My own treatment (1969*b* and the present work, chapter 2) follows in the Zaitzevian tradition of Hutchinson and Mauer, with modifications, in interpreting the lint-bearing cottons as a limited number of species

It is worth noting that of the two traditions of classification of *Gossypium*, the one exemplified by Todaro, Watt, and Prokhanov, in which the lint-bearing cottons are fragmented into numerous species with minor differences, shows little agreement among the classifications of the different workers. On the other hand, those classifications of *Gossypium* that have placed the lint-bearing cottons into a relatively few, broad taxa (including those of Parlatore, Aliotta, Hutchinson, and Mauer), have been in relatively close agreement with each other. This agreement argues strongly for the naturalness and correctness of the latter viewpoint. The modern taxonomic treatments of *Gossypium* of Hutchinson and of Mauer have found general acceptance because they have been found useful. They have presented an analysis of the variation pattern in *Gossypium* that has been readily grasped and which closely reflects the experience of others in dealing with these plants.

An additional question concerns the breadth of conception of the genus *Gossypium*; some students of the genus include only the lint-bearing species, and others add some or all of the wild relatives. These ideas have fluctuated back and forth without a clear historical trend, except that contemporary opinion universally accepts the broad, inclusive conception of the genus.

Cienfuegosia

Cienfuegosia was described by Cavanilles in 1787 encompassing a single species, *C. digitata*. This generic name was shortened by A. L. de Jussieu in 1789 to *Fugosia*. Although the latter name is incorrect (merely an orthographic variant, albeit an extreme one), the genus was known by that name over a major portion of the intervening years.

Garcke (1860) was the first to treat the species of *Cienfuegosia* as a coherent group, but only a relatively few species had been described at that time. Hutchinson (1947*a*) published the next significant study of the genus, which included only eleven species, and my own revision (1969*c*, 1974), which is the basis for the treatment given in chapter 2, recognizes twenty-six species.

Over the years there have been varying interpretations of the limits of the genus. Several species that belong in other genera, especially in *Gossypium* and *Alyogyne*, have been included in *Cienfuegosia* at one time or another. This broader and vaguer delimitation of the genus was narrowed by Hutchinson (1947*a*) to a more natural conception, essentially the same delimitation that is accepted here. There remains a question (discussed in more detail in chapter 7) whether *Cienfuegosia* might deserve to be divided into two taxa of generic rank, but additional data are needed to resolve this question.

Thespesia

A critical monograph of *Thespesia* sens. lat. has yet to be written, but several revisionary studies provide guidance to an understanding of the genus. The genus was conceived and segregated by Duhamel, and the name *Bupariti* was published in 1760. Duhamel's name, however, was a nomen nudum, and Solander's later name, *Thespesia*, has been conserved over it. (Fosberg and Sachet [1972] note that the name *Thespesia* is of somewhat doubtful legitimacy, resting principally on its having been conserved.) Both generic names were based on the Linnaean species *Hibiscus populneus*. Since the segregation and establishment of this genus, however, a number of other species of *Thespesia* have been described, and the present study recognizes seventeen species.

Baker (1897) published a synopsis of *Thespesia*, recognizing nine species and accepting a relatively broad circumscription of the genus. Hutchinson's (1947*a*) conception of *Thespesia* was of comparable breadth to that of Baker, although he treated certain of the species in different ways.

On the other hand, Howard (1949) and Exell and Hillcoat (1954) sharply narrowed the limits of *Thespesia* and segregated a number of species out of the genus as the monotypic genera *Montezuma* DC., *Ulbrichia* Urban, *Atkinsia* Howard, and *Thespesiopsis* Exell & Hillc.; they also combined the monotypic genera *Azanza* Alef. and *Shantzia* Lewt. into a genus of two species. This left *Thespesia* with an unstated but small number of species and a very narrow circumscription.

More recently van Borssum Waalkes (1966) described several new species of *Thespesia* from New Guinea which have broadened our knowledge of the genus, and he implied that he regarded an inclusive view of the genus to be preferable. I also made this view explicit (Fryxell, 1968a) while making some necessary nomenclatural changes, but I expressed it only in outline form. As a consequence of this history, the taxonomic treatment of *Thespesia* presented in chapter 2 is the most complete and detailed yet to be published. Even so, it does not claim to be the critical monograph of the genus that is needed, but only a preliminary effort in that direction.

Hampea

THE history of our knowledge of *Hampea* may be described briefly. The genus was established for a single species by Schlechtendal in 1837. Additional species were added in subsequent years, and this information was summarized in a brief revision published by Standley (1927). Little further attention was given the genus until I published a more detailed revision that recognized sixteen species (Fryxell, 1969d), to which four other species have since been added.

Few questions have ever been raised about the circumscription of *Hampea*, since the genus is clearly defined both morphologically and geographically. However, there have been divergent opinions expressed whether the genus should be placed in the Malvaceae or the Bombacaceae. I have reviewed the reasons for this divergence (Fryxell, 1968a), and it is now clear that *Hampea* belongs in the Malvaceae, specifically in the tribe Gossypieae.

Kokia

THE genus *Kokia* was described by Lewton (1912) for a group of species endemic to Hawaii. It has subsequently been discussed by Degener

(1934) and by Hutchinson (1947a). Four species are included, but at least one is extinct and another is known from only a few trees in cultivation. The entire genus is threatened with extinction.

Cephalohibiscus

ULBRICH (1935) described a distinctive new species from New Guinea and erected the genus *Cephalohibiscus* to accommodate it. The genus is monotypic as currently understood. Van Borssum Waalkes (1966) transferred this species to *Thespesia*, but there seems to be ample reason (Fryxell, 1968a) for separating *Cephalohibiscus* as distinct.

Gossypioides

Two species from East Africa and Madagascar that had been known as species of *Gossypium* were placed in the separate genus *Gossypioides* by Skovsted (in Hutchinson, 1947a). Their overall resemblance to *Gossypium* is considerable, but certain morphological characters distinguish the two genera, and they have different chromosome numbers and are genetically isolated. This disposition thus has generally been accepted by subsequent workers.

Lebronnecia

A new species from the Marquesas Islands was found to share certain characters with other genera of the Gossypieae but to be placeable in none of them. Therefore, the new genus *Lebronnecia* (Fosberg and Sachet, 1966) was described to accommodate the new plant. It is known only from a few plants on two islands of the Marquesas group, although a few plants have since been established elsewhere in cultivation.

This historical review shows that a major portion of taxonomic attention has been given to the genus *Gossypium* because of its economic importance and evolutionary interest, and relatively less attention has been given to the other genera of the tribe. It is hoped that this present comparative study will help to redress that imbalance and to put our understanding of these plants on a more even footing.

2

Systematics

Comme on l'a souvent remarque il faudrait une bonne mono-graphie du genre Gossypium, *mais ce ne serait pas l'oeuvre de quelques annes ni d'un botaniste ordinaire.*

Joseph Decaisne

The cotton breeders of the world, for practical reasons if for no other, have had to take an intense interest in the classification of all cottons, whether they seemed to be directly concerned with cultivated varieties or not. As a result, the classification of wild and cultivated cotton is more truly critical and on a more global basis than for any other world crop.

Edgar Anderson, 1969

THIS chapter is an integral part of this natural history and provides the framework for much of the subsequent discussion. It may also stand in its own right as a taxonomic reference to the tribe. Much of the information included here is based upon or summarized from earlier studies of individual genera or other portions of the tribe, as noted in chapter 1. Other parts of the following account are new, especially the treatment of the genus *Thespesia*, or updated. But even though relatively little new material is introduced, there is positive value in bringing together previously available material into a single coherent account so that relationships are made more clearly evident.

Tribe Gossypieae Alefeld

Gossypieae Alef., Bot. Zeit. 19:301. 1861 (as Gossypiidae); Fryxell, Bot. Gaz. 129:301. 1968.

Unarmed tropical and subtropical trees, shrubs, and perennial herbs; herbage more or less punctate with metabolic capacity to synthesize gossypol and related terpenoids. Vestiture generally of stellate hairs or lepidote scales. Leaves alternate, simple, entire, dentate, or lobed, usually with nectaries on principal veins dorsally; foliar nervation usually palmate (or sometimes pedate), rarely pinnate. Stipules subulate, falcate, or auriculate, occasionally foliaceous, caducous or persistent, rarely absent. Flowers regular, mostly perfect (sometimes unisexual), usually large and showy and lasting one day, borne singly or in fascicles in the leaf axils, or on lateral sympodial inflorescences. Pedicels often surmounted by trimerous involucellar nectaries immediately below the involucel. Bractlets of the involucel subulate and inconspicuous to broadly cordate and prominently foliose, enclosing bud; bractlets often 3, sometimes more numerous, occasionally deciduous, sometimes absent. Calyx gamophyllous, deeply lobed to truncate, pentamerous (although pentamerous nature often obscured in truncate calyces), nectariferous within at base. Petals 5, distinct, adnate to staminal column, white, yellow, rose-colored, or purple, often fading rose or red, with or without a maroon spot on claw. Stamens monadelphous, numerous; staminal column pallid or maroon, more or less elongate, surrounding the style, antheriferous in upper portion or throughout length, surmounted by 5 sterile teeth; filaments emerging from column singly or in pairs (rarely more complexly associated); anthers reniform, one-celled, dorsifixed; pollen spheroidal, spinose, with 3–many apertures, usually yellow or cream-colored. Ovary superior, 3- to 5-celled; placentation axile. Style single; stigma 3- to 5-lobed, more or less exserted from androecium, decurrent or capitate, pallid greenish or dark maroon. Fruit a capsule, 3- to 5-celled, dehiscent or sometimes indehiscent, chartaceous, coriaceous, or sometimes ligneous. Seeds 1–many per locule, comose or glabrous, turbinate, exalbuminous, sometimes (especially in *Hampea*) arillate; embryos with conduplicate cotyledons; cotyledons folded, often complexly so, enclosing the mesocotyl and hypocotyl (except in *Cephalohibiscus*). Pigment glands commonly present in embryo. Chromosome numbers: $2n = 20, 22, 24, 26, 52$.

Key to the Genera of the Tribe Gossypieae

A Involucel absent; prostrate perennial herbs; chromosome
  number $2n = 20$ (South America)... *Cienfuegosia* (p. 21)
  Involucel present................................... B
B Bracts of the involucel broad and foliaceous, trimerous C
  Bracts of the involucel subulate or ligulate, 3 to many ..... E
C Foliar and involucellar nectaries lacking; capsule woody;
  bract margins sinuate or irregularly lobed; trees; chromo-
  some number $2n = 24$ (Hawaii) .... *Kokia* (p. 79)

Foliar and involucellar nectaries usually present; capsule coriaceous; bract margins laciniate, dentate, or entire, but not irregularly lobed; usually shrubs .... D

D Stems 5-angled or -winged; style divided apically, chromosome number $2n = 24$ (Africa)... *Gossypioides* (p. 36)

Stems terete or weakly angled; style undivided; chromosome number $2n = 26$ or 52 .... *Gossypium* (p. 37)

E Seeds prominently arillate, glabrous; trees, most species dioecious; chromosome number $2n = 26$ (Mexico to South America) .... *Hampea* (p. 72)

Seeds exarillate, glabrous or comose; perfect-flowered herbs, shrubs, or trees (if flowers unisexual, then herbaceous) .... F

F Seeds reniform with patent hairs; involucral bracts 6; capsule 5-celled, elongate-obovoid; large tree (New Guinea) .... *Cephalohibiscus* (p. 19)

Seeds turbinate; capsule 3- to 5-celled, usually not elongate .... G

G Involucellar bracts 3; trees or shrubs .... H

Involucellar bracts 9 or more .... J

H Capsules 5-celled .... *Thespesia* (p. 84)

Capsules 3-celled..................................... I

I Capsules woody; seeds 1 per locule (Marquesas Islands) .... *Lebronnecia* (p. 83)

Capsules coriaceous; seeds several per locule; chromosome number $2n = 26$.................. *Gossypium* (p. 37)

J Plants herbaceous perennials or subshrubs; leaves entire or dentate; chromosome number $2n = 20$ or 22 .... *Cienfuegosia* (p. 21)

Plants large shrubs or trees; leaves entire; chromosome number $2n = 26$ .... *Thespesia* (p. 84)

Genus *Cephalohibiscus* Ulbrich

Cephalohibiscus Ulbrich, Notizbl. Bot. Gart. Berlin-Dahlem 12:495. 1935.

Trees, up to 30 m tall. Leaves truncate to cordate, 3-lobed (or sometimes simple, ovate), acute or acuminate, 5-nerved (nerves raised above *and* below), sparsely puberulent to glabrate above, sparsely to densely and minutely rusty-puberulent below, obscurely but abundantly punctate. Foliar nectary usually single, basal, sunken, 8–24 mm long; two lateral nectaries rarely present. Petioles punctate, puberulent to glabrate. Stipules linear to falcate or oblanceolate, 5–15 mm long, 0.5–2.0 mm wide, rusty-puberulent, usually persistent. Peduncles axillary, jointed, becoming stout and woody in fruit, often with a reduced (simple) leaf at articulation, commensurate with subtending petiole; pedicels often paired, 1–3 cm long, puberulent to glabrate, usually surmounted by 3 involucral nectaries which may be large (2–4 mm × 1.0–1.5 mm) or obscure or even absent. Involucel of 6 subulate bracteoles, inserted in pairs above each involucellar nectary; bracteoles falcate, 5–10 mm long, 1 mm wide, rusty-puberulent, caducous before anthesis leaving scars; the bracteoles or their scars and the involucellar nectaries (when present) often somewhat irregularly inserted. Calyx globular in bud, narrowly campanulate in flower and fruit, truncate, 12–18 mm long, minutely and densely puberulent to glabrate, becoming ligneous in fruit. Corolla pale brown-pubescent without where exposed in bud, otherwise white (fading pink) and glabrous; petals 2.5–3.0 cm long, prominently nigro-punctate, contorted. Androecial column 2 cm long, white, sparsely punctate, stellate-pubescent (at least below), antheriferous only near apex, terminating in a crown of 5 sterile teeth; anther mass globose; filaments pinkish, flattened or winged; anthers maroon without, white within; pollen yellowish, spheroidal, spinose. Style single, included within androecial column; stigma capitate, 5-lobed, barely exceeding staminal column, immediately surrounded by crown of sterile teeth at apex of staminal column. Capsules 5-loculed, 4.0–5.5 cm long, 10–25 mm at widest, clavate or obovoid-stipitate, obtuse-rostrate, woody, dehiscent, densely covered with fimbriate scales when young, becoming glabrate and verrucate-punctate at maturity; inner carpel wall glabrous, shiny white. Seeds reniform, densely comose; seed hairs red brown, straight and patent, 10–15 mm long. Embryos bent, punctate. Chromosome number not known.

Type species: *Cephalohibiscus peekelii* Ulbrich.

Genus monotypic.

1. *Cephalohibiscus peekelii* Ulbrich, Notizbl. Bot. Gart. Berlin-Dahlem 12:495. 1935. [Fig. 6]

Hibiscus peekelii Ulbrich, *op. cit.*, 494, nomen provis.

Thespesia peekelii (Ulbrich) Borssum Waalkes, Blumea 14:118. 1966.

Characters of the genus.

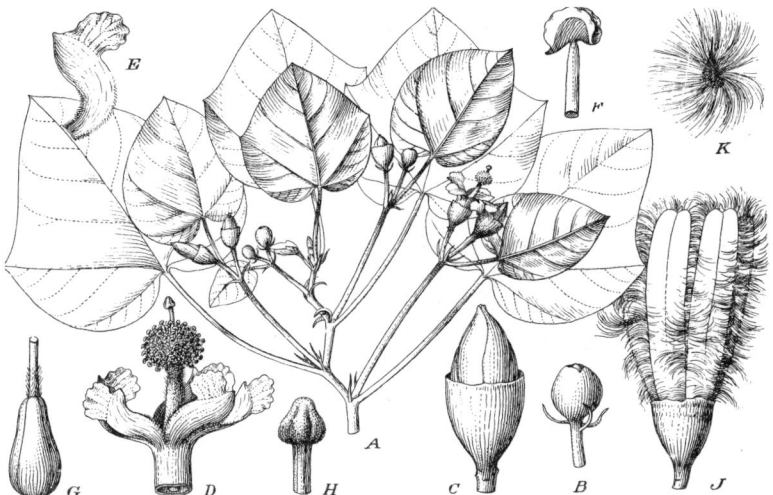

Fig 6. *Cephalohibiscus peekelii*. *A*, branch with four peduncles; *B*, young flower bud with involucel; *C*, older flower bud, the involucellar bracts fallen; *D*, corolla with androecium and style; *E*, single petal; *F*, stamen; *G*, ovary; *H*, stigma; *J*, fruit; *K*, seed. (Reprinted from Ulbrich, Notizbl. Bot. Gart. Berlin-Dahlem 12: 497, pl. 6. 1935. Original drawing by J. Pohl.)

Record (1935) describes the wood structure of this species. Fryxell (1968*a*) discusses the systematic position of the genus and the reasons for maintaining it as distinct from *Thespesia*.

Illustrations: Ulbrich, Notizbl. Bot. Gart. Berlin-Dahlem 12:497, pl. 6; Fryxell, 1968*a*, fig. 6*d*.

Distribution: New Guinea and the Solomon Islands.

Genus *Cienfuegosia* Cavanilles

Cienfuegosia Cavanilles, Diss. 3:174, t. 72, f. 2. 1787; Garcke, Bonpl. 8:148. 1860; Ulbrich, Bot. Jahrb. 50 (suppl.):357. 1914; Hutchinson, New Phytol. 46:125. 1947; Fryxell, Ann. Missouri Bot. Gard. 56:179. 1969; 61:491–493. 1974.

Fugosia Jussieu, Gen. Pl. 274. 1789.

Redutea Ventenat, Hort. Cels, t. 11. 1800.

Cienfuegia Willdenow in L., Sp. Pl., ed. IV, 3:723. 1803.

Elidurandia Buckley, Proc. Acad. Sci. Phila. 1861:450. 1862.

Perennial herbs, subshrubs, or sometimes shrubs with herbaceous to woody stems arising from a woody rootstock. Stems procumbent to erect, glabrous to pubescent, commonly angled, usually punctate. Leaves entire or coarsely serrate, simple to variously divided, sometimes deeply and secondarily so; linear, elliptic, cuneiform, reniform, or divided digitately; pubescent to glabrous, sometimes punctate; foliar nectaries present or absent. Petioles often canaliculate. Stipules large and foliaceous (sometimes clasping) to minute and subulate; caducous or persistent. Peduncles axillary; usually uniflorate although sometimes multiflorate, with or without articulation; trimerous involucellar nectaries present or absent. Involucel sometimes lacking (sect. *Friesia*) but usually present, persistent, and 9- to 12-foliolate (or rarely 3-foliolate by coalescence); bractlets minute and subulate or prominent and linear to spatulate and equaling the calyx. Calyx 5-lobed, usually punctate, glabrous to densely pubescent, sometimes costulate; lobes long-acuminate to rotund-apiculate. Petals sometimes punctate, stellate-pubescent without, white, yellow, pink, or purple, with dark basal spot (sometimes lacking). Staminal column yellowish or dark red; anthers dark red, orange, or yellowish; anther mass globose, oblate, elliptic, or obovate. Style single, sometimes divided apically, usually elongate so that the stigma exceeds the androecium; stigmas 3 (to 5) -lobed (or if style divided, stigmas distinct), decurrent or capitate, dark red or yellowish, sometimes pubescent. Capsule usually 3-loculed (sometimes 4- to 5-loculed), ovoid, globose, or notably elongated; glabrous, stellate-pubescent, or ascending-sericeous, sometimes with hairs on inner margin of suture. Seeds free, 3–8 mm long, turbinate, usually densely comose (rarely subglabrous); seed hairs appressed or patent, up to 10 mm long. Embryos punctate or epunctate with conduplicate cotyledons. Chromosome numbers: $2n = 20, 22$.

Type species: *Cienfuegosia digitata* Cavanilles.

Key to the Species of *Cienfuegosia*

A   Foliar nectaries usually present (rarely suppressed); peduncles articulate; stigmas decurrent; involucel usually multi- (9-)partite but clearly trimerous; chromosome number (where known): $2n = 22$ (subgenus *Articulata*) . . . .   B

Foliar nectaries absent; peduncles nonarticulate; stigmas decurrent or capitate; involucel either absent or multipartite, but not disposed in three groups; chromosome number (where known): $2n = 20$ (subgenus *Cienfuegosia*)

. . . .   I

B   Shrubs; leaves markedly cordate, entire, palmately trilobed with rounded lobes; stipules large, foliose, oblong, ses-

sile; pedicels subtended by foliar leaf (section *Articulata*)
.... 1. *C. gerrardii*
Subshrubs or perennial herbs; leaves cuneate, entire, or
dentate, sometimes shallowly lobed or deeply dissected;
stipules triangular, subulate, or auriculate; pedicels subtended by small bracts C

C   Leaves obovate, entire (sometimes apically tridentate);
flowers dimorphic, in short racemes appearing before
the leaves; stipules subulate; bractlets of involucel 3,
irregular (section *Dioica*)........... 7. *C. heteroclada*
Leaves cuneiform to flabelliform or reniform, entire, dentate, or dissected; penducles axillary, 1- to 2-flowered;
flowers perfect; stipules triangular, subulate, or auriculate; bractlets of the involucel 9 (section *Garckea*, except
C. humbertiana, insertae sedis) .... D

D   Stipules prominent, auriculate-clasping; foliar nectaries 3,
basal; petals yellow; anthers orange ....
.... 2. *C. hildebrandtii*
Stipules inconspicuous, subulate; foliar nectaries 1–3
(rarely absent), medially positioned; petals white, yellow,
or purple; anthers pallid or purple .... E

E   Petioles at least as long as lamina, puberulent or scabrous
especially at distal end; petals white, cream, or yellow..... F
Petioles shorter than the lamina, glabrate; petals purple ... H

F   Petals with maroon spot at base; herbage lacking uncinate
hairs... G
Petals white with no basal spot; herbage covered with
minute uncinate hairs .... 26. *C. humbertiana*

G   Leaves reniform, denticulate, sparsely scabrous below
and on petiole; peduncle exceeding leaves, articulated
near center; petals cream with maroon spot covering
lower third of petal .... 3. *C. welshii*
Leaves flabelliform, slightly 3-lobed to deeply dissected,
dentate, minutely puberulent on margin and petiole;
peduncle shorter than the leaves, articulated basally;
petals bright yellow with maroon spot covering lower
half of petal .... 4. *C. somaliana*

H   Leaves obovate; calyx ecostulate, prominently punctate;
foliar nectaries lacking .... 5. *C. chiarugii*

Leaves cuneiform; calyx costulate, often epunctate; foliar
nectaries 1–3 .... 6. *C. hearnii*

I Involucellar nectaries usually present; capsule epunctate
with inner suture hairs present; bractlets of the involucel
subulate, usually much shorter than the calyx (section
Cienfuegosia) .... J

Involucellar nectaries absent; capsules usually punctate,
without inner suture hairs (except *C. hitchcockii*);
bractlets of involucel lanceolate to spatulate and subequal to calyx or totally absent .... O

J Leaves deeply digitately divided, often secondarily so; petal
spot (rarely lacking) and androecial column carmine ...
.... 8. *C. digitata*

Leaves entire, lobed, rarely with secondary divisions as
above, or trifoliolate; petal spot and androecial column
dark maroon (or spot absent and column pallid)......... K

K Leaves very narrow, lanceolate to oblong, simple (lower
leaves rarely trilobed); involucellar nectaries usually
lacking; petal spot absent; pollen yellow ....
.... 9. *C. yucatanensis*

Leaves broad and entire or variously divided; involucellar
nectaries present; petal spot present; pollen orange or
reddish L

L Bractlets of the involucel 3–10 mm long; leaves simple
to 3-lobed, apiculate .... 10. *C. rosei*

Bractlets of the involucel 1–3 mm long; leaves simple to
3-lobed or narrowly trifoliolate, acute to obtuse........ M

M Stigma decurrent, largely included within the androecium
.... 11. *C. heterophylla*

Style and stigma greatly exceeding the androecium, usually
subequal to the petals; stigma often subcapitate .... N

N Leaves trifoliolate, leaflets narrow; petals yellow with large
basal spot having yellow radii ...... 12. *C. subternata*

Leaves entire, cuneiform to variously divided (sometimes
secondarily so); petals white or pale yellow with small
basal spot usually lacking radii .... 13. *C. tripartita*

O Stipules linear or filiform; involucel subequal to calyx;
petals punctate...................................... P

Stipules subulate or auriculate; involucel subequal to calyx

or lacking; petals epunctate or at most obscurely
punctate .... S

P Involucellar bracts manifestly spatulate-obovate, acuminate; capsules ciliate on internal suture margins; leaves
trilobed, up to 8 cm long (section *Spathulata*) ....
.... 17. *C. hitchcockii*
Involucellar bracts linear-lanceolate to weakly spatulate;
capsules glabrous internally; leaves elliptic or if trilobed,
less than 3 cm long (section *Robusta*) .... Q

Q Leaves penninerved, ovate to elliptic, unlobed, short-
petioled; petals 2.5–5 cm long .... R
Leaves palminerved, trilobed, 1.5–2 cm long; petiole
more than half length of leaf blade; petals 2 cm long .....
.... 16. *C. intermedia*

R Plant more or less pubescent; capsule antrorsely villous;
seeds subglabrous to minutely puberulent ....
.... 14. *C. affinis*
Plant glabrous; capsule glabrous or apically villous; seeds
hairy .... 15. *C. glabrifolia*

S Involucel subequal to calyx; stipules subulate, symmetrical;
plants procumbent or ascending (section *Paraguayana*).... T
Involucel lacking; stipules auriculate, asymmetrical; plants
procumbent (section *Friesia*) .... W

T Leaves dentate, minutely puberulent or pubescent;
petals yellow, sometimes turning blue green in sicco .... U
Leaves entire, glabrous; petals yellow V

U Leaves as broad as long, notably pubescent; calyx and leaf
margins densely ciliate; stigmatic lobes 3–4, pallid; capsule 3- to 4-celled; seed hairs loosely appressed ....
.... 18. *C. sulfurea*
Leaves longer than broad, sparsely scurfy-puberulent becoming glabrate; calyx and leaf margins not or scarcely
ciliate; stigmatic lobes 4–5, dark red; capsule 4- to
5-celled; seed hairs tightly appressed (the seeds appearing hairless) .... 19. *C. drummondii*

V Leaves glaucous, obovate, apiculate, primarily 3-nerved;
stipules persistent, often equaling petioles; pedicels
long (> 5 cm); calyx and involucel nearly glabrous; fruit
glabrous; petioles short .... 20. *C. integrifolia*

Leaves orbicular to ovate, 5-nerved; stipules caducous, inconspicuous; pedicels short (1–4 cm); calyx and involucel scabrous; fruit antrorsely villous; petioles about ½ length of lamina . . . . 21. *C. subprostrata*

W Leaves weakly lobed or variously parted; leaf margins crenate; tips of calyx lobes often hirsute   X

Leaves simple or deeply dissected; leaf margins serrate; tips of calyx lobes glabrous . . . .   Y

X Leaves almost simple to moderately parted (sometimes deeply and secondarily so), glabrate; stigmas pallid . . . . .
. . . . 22. *C. argentina*

Leaves weakly trilobed, hispid below and on petioles; stigmas dark red . . . . 23. *C. hispida*

Y Leaves simple; stipules prominent (often equaling petioles); petioles ½ length of lamina or less . . . 24. *C. ulmifolia*

Leaves deeply dissected; stipules small; petioles ⅔ length of lamina or more . . . . 25. *C. hasslerana*

Subgenus *Articulata* Fryxell

Articulata Fryxell, Ann. Missouri Bot. Gard. 56:194. 1969.

Type species: *Cienfuegosia gerrardii* (Harvey ex Harvey & Sonder) Hochreutiner.

Section I. *Articulata*

1. *Cienfuegosia gerrardii* (Harvey ex Harvey & Sonder) Hochreutiner, Ann. Cons. Jard. Bot. Genève 6:56. 1902.

Fugosia gerrardi Harvey ex Harvey & Sonder, Fl. Cap. 2:588. 1862.

Thespesia rehmannii Szyszylowicz, Polypet. Thal. Rehmann. 44. 1888.

Distribution: South Africa.

Section II. *Garckea* Fryxell

Garckea Fryxell, Ann. Missouri Bot. Gard. 56:196. 1969.

Type species: *Cienfuegosia hildebrandtii* Garcke.

2. *Cienfuegosia hildebrandtii* Garcke, Eichl. Bot. Jahrb. Bot. Gart. Berlin 2:337. 1883.

Distribution: South and East Africa.

3. *Cienfuegosia welshii* (T. Andersson) Garcke, Eichl. Bot. Jahrb. Bot. Gart. Berlin 2:337. 1883.

Hibiscus welshii T. Andersson, J. Linn. Soc. London 5 (suppl. 1):8. 1860.

Fugosia welshii (T. Andersson) Hochreutiner, Ann. Cons. Jard. Bot. Genève 4:174. 1900.

Fig. 7. *Cienfuegosia hearnii.* (Reprinted from Fryxell, Brittonia 19:33. 1967; Fryxell, 1969c, fig. 18.)

Distribution: near Aden (Hearn, 1968) and in northern Somalia, below 700 m.

4. *Cienfuegosia somaliana* Fryxell, Brittonia 19:33. 1967.
Distribution: in northern Somalia between 900 m and 1,700 m and near Diredawa, Ethiopia.

5. *Cienfuegosia chiarugii* Chiovenda, Fl. Somal. 101. 1929.
Distribution: northeastern Somalia.

6. *Cienfuegosia hearnii* Fryxell, Brittonia 19:33. 1967. [Fig. 7]
Distribution: Arabian peninsula (Hearn, 1968).

Section III. *Dioica* Fryxell
Dioica Fryxell, Ann. Missouri Bot. Gard. 56:203. 1969.
Type species: *Cienfuegosia heteroclada* Sprague.

7. *Cienfuegosia heteroclada* Sprague, Kew Bull. 1907:48. 1907 [Fig. 8]
Fugosia heteroclada Sprague, *op. cit.*, 49, pro syn.
Distribution: West Africa in Ghana and Nigeria.

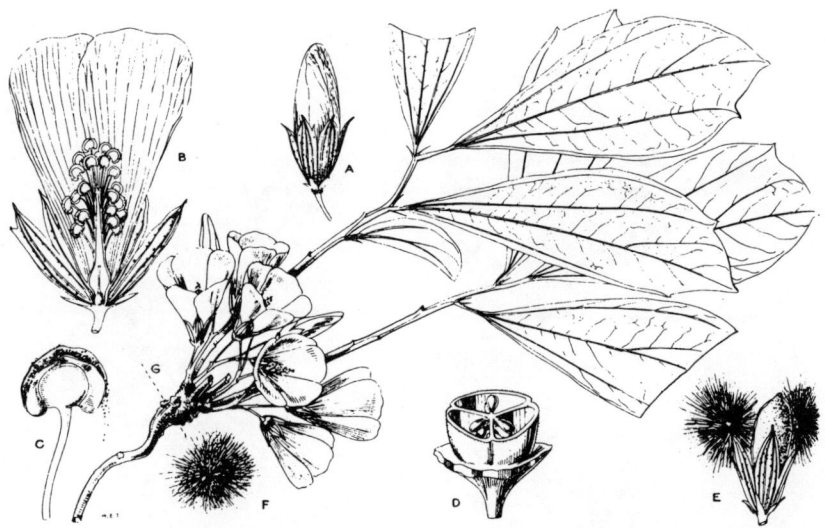

Fig. 8. *Cienfuegosia heteroclada*. A, flower bud; B, longitudinal section of flower; C, stamen; D, cross-section of ovary; E, dehiscence of fruit; F, seed; G, level of soil. (Reprinted from Hutchinson and Dalziel, *Flora of West Tropical Africa*, I, fig. 112. 1928.)

Subgenus *Cienfuegosia*

Section IV. *Cienfuegosia*

8. *Cienfuegosia digitata* Cavanilles, Diss. 3:174. 1787.

Fugosia digitata (Cavanilles) Persoon, Syn. Pl. 2:240. 1806.
Hibiscus cavanillesii Kuntze, Rev. Gen. Pl. 1:69. 1891.
C. digitata var. *lineariloba* Hochreutiner, Ann. Cons. Jard. Bot. Genève 6:56. 1902.
C. junciformis A. Chevalier, Rev. Bot. Appliq. 30:267. 1950.
C. junciformis var. *ruyssenii* A. Chevalier, *op. cit.*, 268.

Distribution: West Africa (Senegal to Nigeria) and southern Africa (Angola to the Transvaal).

9. *Cienfuegosia yucatanensis* Millspaugh, Publ. Field Mus. Nat. Hist. Bot. Ser. 2:74. 1900.

Distribution: Yucatán, Florida Keys, Cuban Cayos, and the Bahamas.

10. *Cienfuegosia rosei* Fryxell, Ann. Missouri Bot. Gard. 56:210. 1969.

Distribution: Mexico, near Tehuantepec, Oaxaca.

11. *Cienfuegosia heterophylla* (Ventenat) Garcke, Bonpl. 8:150. 1860. [Fig. 9]

Redutea heterophylla Ventenat, Hort. Cels, t. 11. 1800.
Fugosia heterophylla (Ventenat) Spach, Hist. Veg. Phan. 3:397. 1834.
Hibiscus redoutei Kuntze, Rev. Gen. Pl. 1:69. 1891.

Fig. 9. *Cienfuegosia heterophylla.* (Reprinted from Ventenat, Hort. Cels, pl. 11. 1800.)

Fugosia punctata Turczaninow, Bull. Soc. Nat. Mosc. 31:196. 1858 (non Cunn. ex Benth.).

Distribution: northern South America and some Caribbean islands.

12. *Cienfuegosia subternata* (Hassler) Fryxell, Taxon 16:321. 1967.

Cienfuegosia heterophylla ssp. *subternata* Hassler, Repert. Sp. Nov. 7:380. 1909.

Distribution: Paraguay.

13. *Cienfuegosia tripartita* (H.B.K.) Gürke in Martius, Fl. Bras. 12(3):578. 1892.

Redoutea tripartita H.B.K., Nov. Gen. Sp. Pl. 5:294.1821.
Fugosia tripartita (H.B.K.) Steudel, Nom., ed. II, 1:649. 1840.
Hibiscus tripartita (H.B.K.) Kuntze, Rev. Gen. Pl. 1:69. 1891.
Fugosia cuneata Bentham, Bot. Voy. Sulph. 68. 1844.
Hibiscus cuneatus (Bentham) Kuntze, *loc. cit.*
Cienfuegosia heterophylla var. *cuneata* (Bentham) Macbride, Publ. Field Mus. Nat. Hist. Bot. Ser. 13:477. 1956.

Distribution: Peru and Ecuador.

Section V. *Robusta* Fryxell
Robusta Fryxell, Ann. Missouri Bot. Gard. 56:216. 1969.

Type species: *Cienfuegosia affinis* (H.B.K.) Hochreutiner.

14. *Cienfuegosia affinis* (H.B.K.) Hochreutiner, Ann. Cons. Jard. Bot. Genève 6:54. 1902. [Fig. 10]

Hibiscus affinis H.B.K., Nov. Gen. Sp. Pl. 5:289. 1821.
Hibiscus sulphureus H.B.K., *loc. cit.*
Cienfuegosia sulphurea (H.B.K.) Hassler, Ostenia 343, 1933.
Hibiscus sulphureus var. *acutifolius* DC., Prodr. 1:451. 1824.
Fugosia lanceolata Jussieu in St.-Hilaire, Fl. Bras. Mer. 1:253. 1825.
Fugosia affinis Jussieu in St.-Hilaire, *loc. cit.*
Hibiscus hilairei Kuntze, Rev. Gen. Pl. 1:69. 1891.
Fugosia phlomidifolia Jussieu in St.-Hilaire, *loc. cit.*
Cienfuegosia phlomidifolia (Jussieu) Garcke, Bonpl. 8:150. 1860.
Hibiscus phlomidifolia (Jussieu) Kuntze, *loc. cit.*
Fugosia campestris Bentham ex Hooker, J. Bot. 4:120. 1842.
Hibiscus campestris (Bentham) Kuntze, *loc. cit.*
Cienfuegosia affinis var. *campestris* (Bentham) Hochreutiner, Ann. Cons. Jard. Bot. Genève 6:54. 1902.
Fugosia guianensis Klotzsch ex Schomburgk, Reise Brit. Guiana 3:1171. 1848.
Fugosia retusa Turczaninow, Bull. Soc. Nat. Mosc. 31:97. 1858.
Cienfuegosia phlomidifolia var. *humilis* Gürke in Martius, Fl. Bras. 12(3):575. 1892.
Cienfuegosia affinis var. *humilis* (Gürke) Hochreutiner, *loc. cit.*
Cienfuegosia riedelii Gürke in Martius, *op. cit.*, 576.
Hibiscus rectiflorus Rusby, Mem. New York Bot. Gard. 7:300. 1927.

Distribution: Venezuela and Brazil to Paraguay and Bolivia.

15. *Cienfuegosia glabrifolia* (St.-Hilaire & Naudin) Blanchard, Ann. Missouri Bot. Gard. 65:766. 1978.

Hibiscus glabrifolius St.-Hilaire & Naudin, Ann. Sci. Nat., Bot., ser. 2, 18:40. 1842.
Cienfuegosia cuyabensis Pilger, Engl. Bot. Jahrb. 30:171. 1902.

Fig. 10. *Cienfuegosia affinis.* 1, petal; 2, staminal tube with exserted style and stigmas; 3, free part of filament with anther; 4, pollen; 5, ovary, with a section of the adnate basal parts of the petals and staminal tube; 6, ovary with one carpel wall removed, showing the ovules in position; 7, dehiscent capsule, the calyx removed; 8, dehiscent capsule with one valve removed, revealing the placental axis and seed; 9, seed; 10, seed, the seed coat longitudinally sectioned, leaving the inner integument with chalaza; 11, seed in longitudinal section; 12, embryo, with the cotyledons slightly expanded and the folds opened to show the radicle. (Reprinted from St.-Hilaire, Fl. Bras. Mer. 1:253, pl. 50. 1825; given there as *Fugosia phlomidifolia.*)

Distribution: central Brazil, in Matto Grosso.

16. *Cienfuegosia intermedia* Fryxell, Brittonia 19:37. 1967.

Distribution: Mexico, exact locality unknown (for discussion, see Southwestern Nat. 18:479–481. 1974).

Section VI. *Spathulata* Blanchard
Spathulata Blanchard, Ann. Missouri Bot. Gard. 65:765. 1978.

17. *Cienfuegosia hitchcockii* (Ulbrich ex Kearney) Blanchard, Ann. Missouri Bot. Gard. 65:764. 1978.
Hibiscus hitchcockii Ulbrich ex Kearney, Leafl. W. Bot. 7:271. 1955.

Distribution: Ecuador and Peru.

Section VII. *Paraguayana* Fryxell
Paraguayana Fryxell, Ann. Missouri Bot. Gard. 56:220. 1969.

Type species: *Cienfuegosia sulfurea* (Jussieu in St.-Hillaire) Garcke.

18. *Cienfuegosia sulfurea* (Jussieu in St.-Hilaire) Garcke, Bonpl. 8:150. 1860. [Fig. 11]
Fugosia sulfurea Jussieu in St.-Hilaire, Fl. Bras. Mer. 1:252. 1825.
Hibiscus jussieui Kuntze, Rev. Gen. Pl. 1:69. 1891.
Cienfuegosia sulphurea var. *genuina* Gürke in Martius, Fl. Bras. 12(3):577. 1892.
Cienfuegosia drummondii (A. Gray) Lewton emend. Hassler, Ostenia 342. 1933, pro parte.
Cienfuegosia drummondii var. *pubescens* Hassler, *op. cit.*, 343.

Distribution: Paraguay and northern Argentina.

19. *Cienfuegosia drummondii* (A. Gray) Lewton, Bull. Torr. Bot. Club 37:473. 1910. [Fig. 12]
Fugosia drummondii A. Gray, Pl. Wright. 1:23. 1852.
Hibiscus drummondii (A. Gray) Kuntze, Rev. Gen. Pl. 1:69. 1891.
Cienfuegosia sulphurea Garcke emend. Robinson in Gray & Robinson, Syn. Fl. N.A. 1(1):337. 1895, pro parte.
Cienfuegosia sulphurea var. *drummondii* (A. Gray) Hochreutiner, Ann. Cons. Jard. Bot. Genève 4:173. 1900.
Elidurandia texana Buckley, Proc. Acad. Sci. Phila. 1861:450. 1862.
Hibiscus pulverulentus Grisebach, Abh. Königlich. Ges. Wiss. Göttingen 24:49. 1879.
Fugosia pulverulenta (Grisebach) Hochreutiner, *loc. cit.*
Cienfuegosia sulphurea var. *glabra* Gürke in Martius, Fl. Bras. 12(3):577. 1892.
Cienfuegosia sulphurea var. *glabra* f. *intermedia* Chodat & Hassler, Bull. Herb. Boiss., ser. 2, 5:302. 1905.
Cienfuegosia sulphurea var. *major* Hassler, Repert. Sp. Nov. 7:379. 1909.
Cienfuegosia drummondii var. *genuina* Hassler, Ostenia 343. 1933.
Fugosia sulphurea var. *trifida* Grisebach ex Rodrigo, Darwiniana 5:220. 1941, nomen nudum.

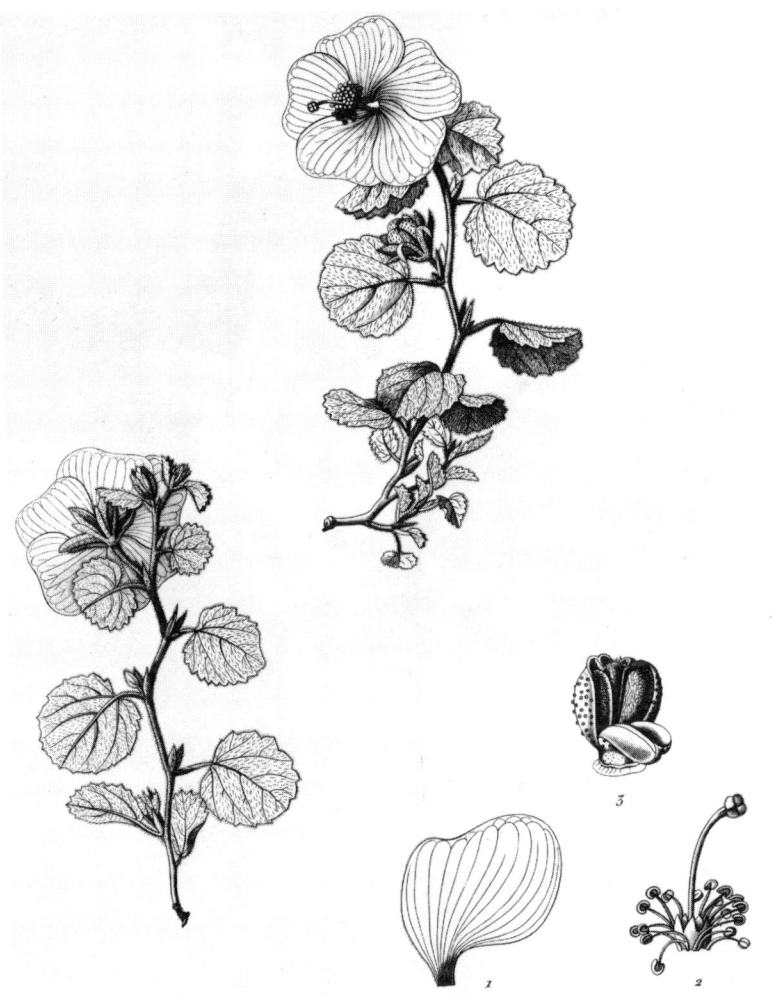

Fig. 11. *Cienfuegosia sulfurea*. 1, petal, inner surface; 2, style and upper part of the staminal tube, transversely cut; 3, capsule (immature). (Reprinted from St.-Hilaire, Fl. Bras. Mer. 1:252, pl. 49. 1825; given there as *Fugosia sulfurea*.)

Distribution: Paraguay and northern Argentina; southern Texas.

20. *Cienfuegosia integrifolia* (Chodat & Hassler) Fryxell, Ann. Missouri Bot. Gard. 56:227. 1969.

Cienfuegosia sulphurea var. *integrifolia* Chodat & Hassler, Bull. Herb. Boiss., ser. 2, 5:302. 1905.

Distribution: Paraguay.

21. *Cienfuegosia subprostrata* Hochreutiner, Ann. Cons. Jard. Bot. Genève 6:57. 1902.

Distribution: central Paraguay.

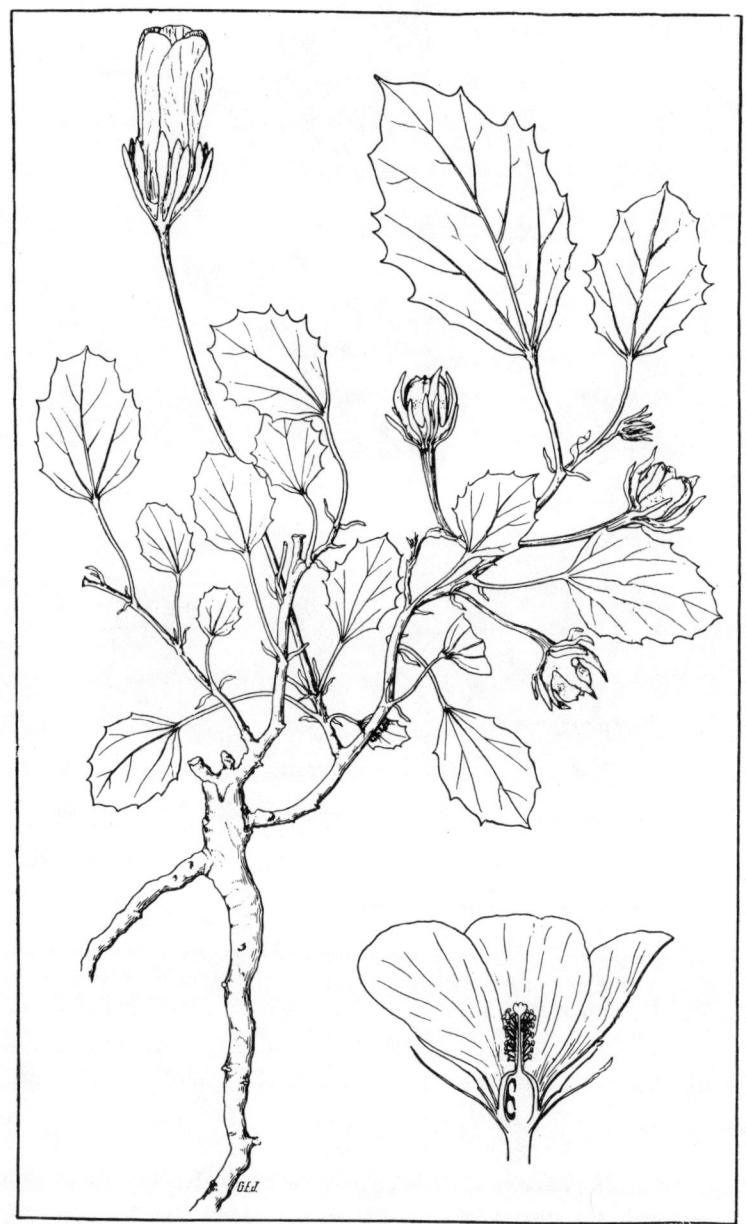

Fig. 12. *Cienfuegosia drummondii.* (Reprinted from J. G. Smith, Ann. Rep. Missouri Bot. Gard. 6:113–120, pl. 49. 1895; given there as *Fugosia drummondii.*)

Fig. 13. *Cienfuegosia ulmifolia*. (Reprinted from Fryxell, Brittonia 19:35. 1967.)

Section VIII. *Friesia* Fryxell

Friesia Fryxell, Ann. Missouri Bot. Gard. 56:228. 1969.

Type species: *Cienfuegosia argentina* Gürke.

22. *Cienfuegosia argentina* Gürke in Martius, Fl. Bras. 12(3):579. 1892.

Hibiscus argentinus (Gürke) Kuntze, Rev. Gen. Pl. 3(2):19. 1898.

Fugosia argentina (Gürke) Hochreutiner, Ann. Cons. Jard. Bot. Genève 4:172. 1900.

Distribution: Bolivia, Paraguay, and northern Argentina.

23. *Cienfuegosia hispida* R. E. Fries, Kungl. Sv. Vet. Akad. Handl. 24:33. 1947.

Distribution: northern Argentina.

24. *Cienfuegosia ulmifolia* Fryxell, Brittonia 19:35. 1967. [Fig. 13]

Distribution: northern Argentina and Paraguay.

25. *Cienfuegosia hasslerana* Hochreutiner ex Chodat & Hassler, Bull. Herb. Boiss., ser. 2, 5:302. 1905.

Cienfuegosia argentina var. *hasslerana* (Hochreutiner) Hassler, Repert. Sp. Nov. 7:381. 1909.

Distribution: northern Argentina and Paraguay.

Incertae sedis:

26. *Cienfuegosia humbertiana* (Hochreutiner) Fryxell, Ann. Missouri

Bot. Gard. 61:492. 1974.
Hibiscus humbertianus Hochreutiner, Candollea 5:9. 1932.
Distribution: Madagascar.

Descriptions, illustrations, distribution maps, type and specimen citations, and other details on *Cienfuegosia* may be found in the monograph of the genus (Fryxell, 1969c); see also Fryxell, 1974, and Blanchard, 1978.

Genus *Gossypioides* Skovsted ex J. B. Hutchinson
Gossypioides Skovsted ex J. B. Hutchinson, New Phytol. 46:131. 1947; Exell in Exell & Wild, Fl. Zamb. 1:432. 1961.

Clambering or scandent shrubs, much branched, up to 3 m tall; herbage pubescent, glabrescent, or glabrous, more or less prominently punctate. Stems 5-angled or -winged. Leaves 3- to 5-lobed, cordate, entire; lobes ovate to lanceolate, acuminate. Foliar nectaries 1–5, on principal veins below. Petioles subcylindrical to sharply quadrangular, ½–⅔ length of lamina. Stipules falcate or auriculate-clasping, usually persistent. Flowers axillary or on 1- to 4-flowered sympodial inflorescences. Pedicels 1–2 cm long, 5- or 6-angled. Involucellar nectaries lacking (?). Bracts of the involucel 3, distinct, ovate-cordate, dentate or laciniate, persistent. Calyx short (6mm), sub-truncate or very shortly 5-lobed or -undulate. Petals 2.5–5.5 cm long, orange-yellow or yellow with dark maroon spot on base. Staminal column bearing numerous anthers on short (1–3 mm) filaments. Style single below (within staminal column) but more or less divided apically into 3–5 lobes; stigmas distinct, clavate, exceeding the androecium by 3–5 mm. Capsule 3- to 5-celled, globose to elongated, with internal suture hairs more or less well developed. Seeds ellipsoid, subglabrous to densely and finely lanate; seed hairs red-brown. Embryos with conduplicate cotyledons, punctate. Chromosome number: $2n = 24$.
Type species: *Gossypioides kirkii* (Masters) J. B. Hutchinson.

Key to the Species of *Gossypioides*
A   Foliar nectaries in center of leaf lobes, distal to sinuses; stipules prominent, auriculate-clasping; bracts of the involucel nearly as broad as long, 10- to 17-dentate; petals 3 cm long, barely exceeding bracteoles; capsule sub-globose, 3- to 4-celled; inner suture hairs copious, filling locule, with 2 seeds embedded in red-brown hairs

at base of each locule; seeds minutely puberulent to subglabrous, lying in but not attached to hairs, which are
capsular in origin .... 1. *G. kirkii*
Foliar nectaries (sometimes single and basal) proximal to
leaf sinuses; stipules inconspicuous, falcate; bracts of the
involucel ca. twice as long as broad, 5- to 9-laciniate;
petals 5 cm long, greatly exceeding bracteoles; capsule
elongated, 5-celled; inner suture hairs sparse, seeds
8–9 per locule, densely lanate; seed hairs fine, red-
brown, 10–12 mm long, surrounding and attached to
seeds .... 2. *G. brevilanatum*

1. *Gossypioides kirkii* (Masters) J. B. Hutchinson, New Phytol.
   46:132. 1947. [Fig. 14]
Gossypium kirkii Masters, J. Linn. Soc. Bot. 19:214. 1882.
Gossypium kirkii ssp. *scandens* Roberty, Candollea 13:31. 1950 (= *Gossypium
    bussei* Gürke inedit.).

Exell (1961, p. 432) drew attention to "two apparently well-
marked varieties: tomentose and glabrous," to which Roberty had earlier given nomenclatural recognition in subspecific rank. Hutchinson
(1947a) noted the existence of a genetic sterility barrier between two
strains of *G. kirkii* which evidently correspond to these two entities:
ssp. *kirkii* (tomentose with angled stems, low growth habit), and ssp.
scandens (glabrous with winged stems, vigorous growth). Although
neither Hutchinson nor Exell noted any correlation of these
morphological differences with geographical distribution, it is possible
that more detailed study of the distribution, morphology, and genetics
of these plants will reveal the existence of two distinct taxa.
Illustrations: Exell, 1961, pl. 87; Watt, 1907, pl. 51.
Distribution: East coast of Africa, from Kenya to Natal.

2. *Gossypioides brevilanatum* (Hochreutiner) J. B. Hutchinson, New
   Phytol. 46:132. 1947. [Fig. 15]
Gossypium brevilanatum Hochreutiner, Candollea 2:140. 1925.
Gossypium kirkii ssp. *brevilanatum* (Hochreutiner) Roberty, Candollea 13:31.
    1950.
Illustration: Hochreutiner, 1955, pl. 32, figs. 5–8.
Distribution: Madagascar.

Genus *Gossypium* Linnaeus
Gossypium L., Sp. Pl. 693. 1753. Todaro, Relaz. Cult. Cot. (1877); Watt, Wild
    and Cult. Cot. Plants of the World (1907); J. B. Hutchinson, Evol. Gos-

Fig. 14. *Gossypioides kirkii*. 1, flowering shoot with persistent stipules, numerous glands, and woolly bracteoles; 2, bud showing internal glands; 3, calyx opened out showing whorl of hairs; 4, ripe fruit and subtended bracteole; 5, seed, natural size, showing floss; 6, seed with floss removed, a portion adhering to the apex; 7, seed enlarged, showing smooth, polished surface and lighter-colored bands. Note that the sharp angles on the stems and petioles are not shown in this drawing. (Reprinted from Watt, 1907, pl. 51; given there as *Gossypium kirkii*.)

Fig. 15. *Gossypioides brevilanatum*. 5, flowering branch (× 2/3); 6, calyx (× 3/2); 7, staminal column and styles (× 3/2); 8, fruit (× 2/3). (Reprinted from Hochreutiner, 1955, pl. 32; given there as *Gossypium brevilanatum*.)

syp. (1947); Prokhanov, Bot. Zhur. 32:61. 1947: Wouters, Publ. Inst. Nat. Étude Agron. Congo Belge no. 34. 1948; Roberty, Candollea 9:19. 1942; 10:345. 1946; 13:9. 1950; Mauer, Proiskhozhd. i Sist. Khlopchat. [Orig. and Syst. of Cott.](1954); Fryxell, Adv. Fr. Pl. Sci. 10:31. 1965; Fryxell, Taxon 18:585. 1969; Fryxell, U.S.D.A. Tech. Bull. 1491. 1976.

Xylon Miller, Gard. Dict. Abridg., ed. IV, 3. 1754.

Ingenhouzia Mociño & Sessé ex DC., Prodr. 1:474. 1824. (non *Ingenhoussia* Dennst. 1818).

Sturtia R. Brown in Sturt, Exped. Cen. Austral. 2:app. 68. 1849.

Thurberia A. Gray, Mem. Amer. Acad. Arts, n.s., 5:308. 1854 (non Bentham, 1881).

Erioxylum Rose & Standley, Contr. U.S. Natl. Herb. 13:307. 1911.

Selera Ulbrich, Verh. Bot. Vereins Prov. Brandenburg 55:50. 1913.

Notoxylinon Lewton, J. Wash. Acad. Sci. 5:305. 1915, pro parte.
Neogossypium Wouters (pro sect.) ex Roberty, Coton Fibres Trop. 4:89. 1949.

Shrubs or subshrubs, erect or decumbent, or small to medium-sized trees, often deciduous, usually punctate throughout. Foliage glaucous or glabrate to puberulent or hirsute. Leaves ovate-cordate, entire, acuminate, shallowly lobulate, or deeply palmately 3- to 7-lobed (actually trifoliolate in one species), palmately or pedately 3- to 7-nerved. Foliar nectaries 1–5, rarely absent, colorless or bright red, basally or distally placed, sometimes elongated. Petioles terete or quadrangular. Stipules filiform, subulate, or falcate, persistent or caducous. Flowers borne singly or clustered on axillary pedicels or severally on sympodially branching lateral inflorescences; pedicels usually surmounted by trimerous involucellar nectaries. Involucel of three bracts inserted above the nectaries. Bracts usually distinct, filiform to subulate to broadly cordate; much shorter than calyx to nearly equaling corolla and enclosing fruit; entire, dentate, or laciniate; usually persistent in fruit, but sometimes caducous at anthesis. Calyx truncate, 5-toothed, or deeply 5-lobed, often prominently punctate. Petals often large and showy; white, cream yellow, rose, or mauve; usually with a large dark spot on claw (spot sometimes reduced or completely absent); usually pubescent without, glabrous within. Staminal column usually glabrous and pallid but sometimes darkly pigmented (by extension of petal spot), antheriferous throughout length or only in upper half; pollen yellow or cream-colored. Style and stigma usually long and slender (often greatly exceeding androecium, but sometimes shorter), clavate; style undivided, pallid; stigma decurrent 3- to 5-lobed. Capsule 3- to 5-celled, glabrous or sometimes minutely puberulent, chartaceous, coriaceous, or ligneous, often prominently punctate, dehiscent and opening slightly or flaring widely. Seeds two to several per locule, usually free but sometimes more or less coherent into a single unit for each locule; usually comose, sometimes minutely puberulent or glabrous; turbinate, sometimes angularly so, 4–12 mm long; seed hairs white to various shades of brown, copious, 1–3 cm long, usually crimped but patent in two species. Embryos with conduplicate cotyledons, punctate or epunctate. Chromosome numbers: $2n = 26, 52$.

Type species: *Gossypium arboreum* Linnaeus.

Key to the Species of *Gossypium*

A Embryos with no (or *very* few) gossypol glands (although these sometimes become evident in cotyledons following germination); corolla mauve or white; erect or decumbent shrubs or subshrubs; wild species from Australia (except *G. triphyllum* from South-West Africa, now Namibia); chromosome number: $2n = 26$ (subgenus *Sturtia*) .... **B**

    Embryos with prominent gossypol glands (often black, sometimes reddish, or even translucent yellow, but always prominent); corolla cream, yellow, or rose; erect shrubs or trees (rarely decumbent or scandent subshrubs); chromosome numbers: $2n = 26$ or 52 .... **M**

B Erect shrubs; herbage odoriferous when crushed, glaucous, with raised black tubercles on stems and petioles; bracts of the involucel broadly cordate to narrowly triangular, exceeding calyx; calyx truncate to short-toothed; petals mauve; cotyledons obcordate (sometimes nearly bifid) (section *Sturtia*) .... **C**

    Erect or decumbent subshrubs; herbage nonodoriferous, glabrate to puberulent or pubescent, punctate but lacking raised tubercles on stems and petioles; bracts of the involucel linear-subulate to filiform, usually shorter than the calyx; calyx 5-lobed; petals mauve or white; cotyledons obtuse or emarginate............... **E**

C Leaves simple or moderately 3-lobed; foliar nectaries 1–3, obscure, basal; bracts of the involucel ovate to broadly cordate; seeds 3–4 mm long; much-branched shrubs, as broad as tall .... **D**

    Leaves deeply 3- to 5-lobed, the lobes ovate-lanceolate, acuminate; foliar nectaries 3–5, often red, prominent, distal; bracts of the involucel triangular to linear-lanceolate (rarely ovate-lanceolate), entire; seeds 4–6 mm long; little-branched shrubs, taller than broad .... 3. *G. robinsonii*

D Leaves simple, often adaxially folded or rolled; bracts of the involucel ovate, entire (rarely laciniate); seed hairs whitish .... 1. *G. sturtianum* var. *sturtianum*

Leaves flat, usually 3-lobed, with auriculate appendages at base; lobes rounded, apiculate; bracts of the involucel broadly cordate, entire or 3- to 5-dentate; seed hairs brownish . . . . 2. *G. sturtianum* var. *nandewarense*

E Erect or decumbent subshrubs; foliage finely tomentose to pilose to glabrate; leaves simple, the foliar nectary obscure; peduncles uniflorate, 1–8 cm long, articulated near or below middle; bracts of the involucel erect or reflexed, subulate, usually shorter than the calyx; calyx lobes usually long and broadly foliaceous; capsules glabrous, 3-loculed; seeds glabrous, sometimes with a small aril (section *Grandicalyx*) . . . .   F

Usually erect subshrubs; foliage finely puberulent to coarsely stellate-pubescent; leaves simple, lobed, or deeply parted; foliar nectary red or colorless, sometimes prominent; inflorescences often multiflorate, sympodial; cleistogamy common; pedicels 1–2 cm long; bracts of the involucel erect, linear to filiform, subequal to calyx; calyx lobes narrowly acuminate; capsules puberlent or glabrous, 3- to 5-loculed; seeds comose (section *Hibiscoidea*) . . . .   J

F Foliage finely and densely puberulent; petals white . . . .   G
Foliage coarsely pilose or glabrate; petals mauve . . . .   H

G Peduncle up to 8 cm long, ribbed, articulated well below center; calyx lobes prominently 3-ribbed . . . .
. . . . 4. *G. costulatum*
Peduncle 1–1.5 cm long; calyx lobes 1-nerved . . . .
. . . . 5. *G. pulchellum*

H Leaves elliptic, acute, penninerved, short-petioled; bracts of the involucel erect; plant erect . . . 8. *G. cunninghamii*
Leaves cordate, acuminate, palmately nerved, long-petioled; bracts of the involucel reflexed   I

I Plant decumbent, glabrate; leaves small . . . .
. . . . 6. *G. populifolium*
Plant erect, pilose; leaves large . . . . 7. *G. pilosum*

J Capsules puberulent; herbage finely puberulent; calyx tube constricted basally; gossypol glands of the calyx few, more frequent on distal margin . . . .   K
Capsules glabrous or nearly so, prominently punctate;

herbage coarsely pubescent; calyx tube rounded basally; gossypol glands of the calyx uniformly distributed ...... **L**

K Leaves trifoliolate; seed hairs crimped, appressed; foliar nectary small, colorless .... 12. *G. triphyllum*

Leaves simple or shallowly lobed; seed hairs straight, patent; foliar nectary usually prominent (4–12 mm long), red .... 9. *G. australe*

L Calyx 6–12 mm long; seed hairs straight, patent; capsule depressed-apiculate, the beak 0.5–2.0 mm long .... 10. *G. nelsonii*

Calyx 18–24 mm long; seed hairs crimped, appressed; capsule ovoid, the beak 3–5 mm long ... 11. *G. bickii*

M Relatively robust shrubs or trees (sometimes smaller shrubs) bearing no commercial cotton fibers; bracts of the involucel entire or sometimes laciniate; seeds comose or subglabrous; wild species of the New World; chromosome number: $2n = 26$ (subgenus *Houzingenia*) .... **N**

Relatively small shrubs and subshrubs (or if arborescent, then bearing cotton); bracts of the involucel commonly incised, sometimes entire; seeds variously comose, sometimes producing commercial cotton; chromosome number $2n = 26$ or 52 .... **Y**

N Small to large shrubs; petals cream or yellow, 1.5–4.5 cm long, with small petal spot (sometimes absent); capsules subglobose or oblong; seeds subglabrous or with seed hairs tightly appressed (section *Houzingenia*) ...... **O**

Large shrubs or trees; petals various shades of rose, 3–8 cm long, with large petal spot (covering lower one-third or one-half of petal); capsules ovoid-elongate; seeds comose (section *Erioxylum*) .... **U**

O Seeds 3–6 mm long, subglabrous; inflorescences sympodial; petal spot small, sometimes vestigial or absent; involucel persistent in fruit .... **P**

Seeds 8–10 mm long, with tightly appressed seed hairs; flowers axillary; petal spot small but well marked; involucel caducous by anthesis; leaves glabrate, simple to shallowly lobed (subsection *Caducibracteolata*) .... **S**

P Petals cream to pale yellow; seeds 3–5 mm long; leaves deeply 3- to 5-lobed; bracts of the involucel usually

entire (if toothed, then linear); foliage glabrate; capsules mostly 3-celled; stems 5-angled or -ridged (subsection *Houzingenia*)............................ Q

Petals yellow; seeds 5–6 mm long; leaves simple or shallowly lobulate; bracts of the involucel laciniate or
dentate, cordate; foliage softly tomentose (rarely
glabrate); capsules 4- to 5-celled; stems terete (subsection
Integrifolia).. R

Q Bracts of the involucel ovate-cordate, entire; shrub to 6 m
tall; stipules falcate, 7–10 mm long; calyx irregularly
5- to 10-toothed; petals 2–3 cm long, pale yellow; capsule 15–18 mm long .... 13. *G. trilobum*

Bracts of the involucel linear, entire or rarely few-toothed
at apex; shrub to 4 m tall; stipules linear-subulate,
5–7 mm long; calyx truncate; petals 1.5–2.5 cm long,
cream; capsule 10–15 mm long .... 14. *G. thurberi*

R Shrub to 4 m tall; petals strongly plicate, 3–4 cm long
with essentially no petal spot; leaves simple; seeds free;
bracts of the involucel 11- to 17-laciniate ....
.... 15. *G. klotzschianum*

Shrub 0.5–1.5 m tall; petals weakly plicate, 2–3 cm long
with variably developed small dark spot at base (rarely
absent); leaves simple or 3-lobulate; seeds often adherent within the locule; bracts of the involucel 7- to
11-laciniate .... 16. *G. davidsonii*

S Peduncle 2–7 cm long with small petiolate leaf at submedian articulation; bracts of the involucel linear, 1–3
mm wide, caducous in bud; leaves unlobed; seed hairs
brownish .... 18. *G. armourianum*

Peduncle 0.5–2 cm long, with subulate caducous bract
at submedian articulation; bracts of the involucel
lanceolate to ovate, 4–9 mm wide, caducous at anthesis;
leaves shallowly 3-lobed .... T

T Bracts of the involucel lanceolate, entire, 4–7 mm wide;
staminal column 7–8 mm long, the filaments 2–3 mm
long; carpel walls 1.5 mm thick; peduncles 0.5–1.5 cm
long; gossypol glands on upper leaf surface obscure; seed
hairs whitish .... 17. *G. harknessii*

Bracts of the involucel ovate, 3- to 7-laciniate, 7–9 mm

wide; staminal column 16–17 mm long, the filaments 5–6 mm long; carpel walls 0.5 mm thick; peduncles ca. 2 cm long; gossypol glands on upper leaf surface manifest; seed hairs brownish . . . . 19. *G. turneri*

U Trees, usually with a single trunk; flowers axillary (sometimes fasciculate), mostly appearing after leaves are shed; bracts of the involucel triangular-subulate, much shorter than calyx (subsection *Erioxylum*)   V

Large subarborescent shrubs usually with multiple trunks; inflorescences sympodial, the flowers appearing when the plants are in full foliage; bracts of the involucel ovate-cordate, enclosing the bud and greatly surpassing the calyx; staminal column often pigmented   X

V Leaves spirally arranged; calyx subtruncate or shallowly (1–3 mm) toothed; flowers usually solitary; seed hairs brown . . . .   W

Leaves distichous, pedately veined, 3-lobed; calyx 5-lobed (lobes 6–10 mm long); flowers usually in fascicles; seed hairs grayish . . . . 22. *G. lobatum*

W Leaves simple, subtruncate, palmately nerved, ovate; capsules 3-celled; foliage softly puberulent to glabrate . . . . 20. *G. aridum*

Leaves usually trilobed, cordate, pedately nerved, capsules 3- to 5-celled; foliage sparsely tomentose to glabrate . . . . 21. *G. laxum*

X Bracts of the involucel entire, marginally connate in bud through interlocking hairs; leaves deeply 3-lobed; foliar and involucellar nectaries lacking; anther mass columnar (subsection *Selera*) . . . . 23. *G. gossypioides*

Bracts of the involucel deeply laciniate or fimbriate, free; leaves simple, cordate; foliar and involucellar nectaries present; anther mass globose to ellipsoid (subsection *Austroamericana*) . . . . 24. *G. raimondii*

Y Leaves simple or lobed (sometimes deeply parted), the lobes commonly narrowed at their base, obtuse or acuminate; bracts of the involucel linear to cordate, entire or variously incised, free or sometimes connate basally; flowers and fruits erect or pendent; seeds comose, bearing short hairs or commercial cotton; wild

species or diploid cultigens of the Old World; chromosome number: $2n = 26$ (subgenus *Gossypium*) . . . .   Z

Leaves lobed (rarely deeply parted), the lobes acuminate,
commonly broadest at the base; bracts of the involucel
cordate, laciniate, free, flowers and fruits erect; seeds
usually bearing commercial cotton; New World tetraploids, mostly cultigens, cosmopolitan in cultivation;
chromosome number: $2n = 52$ (subgenus *Karpas*) . . . . . .   h

Z Erect shrubs or subshrubs, some in cultivation for seed
fibers; leaves 3- to 7-lobed, the lobes rounded (except
very narrow in certain cultigens); seeds comose, bearing
short hairs or commercial cotton (section *Gossypium*) . . . . .   a

Erect, decumbent, or scandent shrubs or subshrubs, none
cultivated; leaves simple or lobed; seeds comose but
bearing no commercial cotton (section *Pseudopambak*) . . .   d

a Cultigens, bearing cotton; bracts of the involucel ovate-
cordate, entire or incised; capsules smooth (subsec-
tion *Gossypium*). . . . .   b

Wild plants, bearing no cotton; bracts of the involucel
linear, more or less divided or toothed apically; cap-
sules verrucose (subsection *Anomala*)   c

b Bracts entire or few-laciniate, enclosing bud, sometimes
basally connate; capsules elongated, flaring widely,
releasing cotton; corolla color variable . . . .
. . . . 25. *G. arboreum*

Bracts 5- to 13-dentate, often flaring, usually free; capsules subglobose, not flaring on dehiscence (cotton
removed with difficulty); corolla yellow. . . . .
. . . . 26. *G. herbaceum*

c Bracts of the involucel linear, entire or apically 3-toothed;
leaves moderately 3- to 5-lobed, the lobes entire; capsules 3-loculed . . . . 27. *G. anomalum*

Bracts of the involucel apically 3-toothed to deeply trifid;
leaves deeply 5- to 7-lobed, the central lobe often
secondarily lobed; capsules 3- to 5-loculed . . . .
. . . . 28. *G. capitis-viridis*

d Bracts of the involucel linear to cordate, more or less
incised; leaves simple or lobed; petal spot manifest; plants
erect to decumbent, in arid zones; flowers sometimes

pendent (subsection *Pseudopambak*) .... e

Bracts of the involucel ovate-cordate, entire; leaves simple, cordate-acuminate; petal spot absent; petals bright yellow; plants scandent in *Acacia* scrub (subsection *Longiloba*) .... 33. *G. longicalyx*

e Plants decumbent; leaves deeply lobed; bracts of the involucel ovate-laciniate .... 29. *G. stocksii*

Plants erect; leaves simple or shallowly lobed; bracts of the involucel various................................. f

f Bracts of the involucel broadly cordate, dentate ....
.... 30. *G. somalense*

Bracts of the involucel much narrower, laciniate.......... g

g Bracts of the involucel lanceolate....... 31. *G. incanum*

Bracts of the involucel linear, subentire ....
.... 32. *G. areysianum*

h Petals bright yellow, shiny; petal spot absent; foliar and involucellar nectaries absent; foliage minutely grayish-puberulent; seed hairs red-brown, not commercially usable; bracts of the involucel < 2 cm long ....
.... 34. *G. tomentosum*

Petals cream to yellow, dull-surfaced; petal spot often present; foliar nectaries present and involucellar nectaries usually present; foliage densely pubescent (rarely puberulent) to glabrate; seed hairs white to various shades of brown, usually providing commercial cotton; bracts of the involucel mostly > 2 cm long i

i Leaves deeply 5-lobed, the lobes narrowly lanceolate; capsules smooth, 3-celled, 1.5 cm long, round-apiculate; herbage glabrate; bracts of the involucel ca. 2 cm long; petals yellowish, 2–3 cm long ..... 35. *G. lanceolatum*

Leaves usually less than half-divided, the lobes broadly ovate; capsules smooth or pitted, 3- to 5-celled, > 2 cm long, rotund to elongate; herbage pubescent to glabrate; bracts of the involucel mostly more than 2 cm long; petals cream or yellow, 2–6 cm long .... j

j Capsules 3- to 5-celled, ovoid or subglobose (rarely elongate), smooth; calyx truncate or with acute lobes or long-acuminate teeth, usually < 6 mm long (excluding teeth); fringe hairs on floral nectary absent; stipules

0.5–1.5 cm long (rarely to 2 cm); leaves 3- to 5-lobed, the central lobe triangular to ovate, usually 1.0–1.5 times as long as broad; bracts of the involucel 3- to 19-laciniate, the teeth triangular and acute, or lanceolate and acuminate in distal portion separated by more or less acute sinuses .... 36. *G. hirsutum*

Capsules usually 3-celled, narrowly ovoid, more or less elongate, pitted; calyx usually truncate, to 10 mm long; fringe hairs on floral nectary present; stipules 1–5 cm long; leaves 3- to 7-lobed, the central lobe ovate to lanceolate, usually more than 1.5 times as long as broad; bracts of the involucel 5- to 17-laciniate, the teeth acuminate from the base separated by rounded sinuses .... k

k Plants mostly cultivated, with large capsules and copious white (sometimes tan) seed hairs (cotton) ....
.... 37. *G. barbadense*

Plants wild, with small capsules and sparse seed-hairs (usually tan or brown)............................. l

l Leaves usually about half-divided; from northeastern Brazil .... 38. *G. mustelinum*

Leaves usually more than half-divided; from Galápagos Islands.......................... 39. *G. darwinii*

Subgenus *Sturtia* (R. Brown) Todaro

Sturtia (R. Brown) Todaro, Giorn. R. Ist. Incoragg. Agric. Arti Manifatture Sicil. 1:35. 1863.

Type species: *Gossypium sturtianum* J. H. Willis.

Section I. *Sturtia*

1. *Gossypium sturtianum* J. H. Willis, Vict. Nat. 64:9. 1947 [Fig. 16]

Sturtia gossypioides R. Brown in Sturt, Exp. Cent. Austral. 2:app. 68. 1849.

Gossypium sturtii F. von Mueller, Fragm. 3:6. 1863.

Gossypium australiense Todaro, Giorn. R. Ist. Incoragg. Agric. Arti Manifatture Sicil. 1:35. 1863.

Hibiscus gossypioides (R. Brown) Kuntze, Rev. Gen. Pl. 1:69. 1891.

Cienfuegosia gossypioides (R. Brown) Hochreutiner, Ann. Cons. Jard. Bot. Genève 6:56. 1902.

Gossypium gossypioides (R. Brown) C. A. Gardner, Enum. Plants Austral. Occid. 79. 1930 (non *G. gossypioides* [Ulbrich] Standley).

Illustrations: Watt, 1907, pl. 2; Hutchinson, 1947b, pl. 1; Saunders,

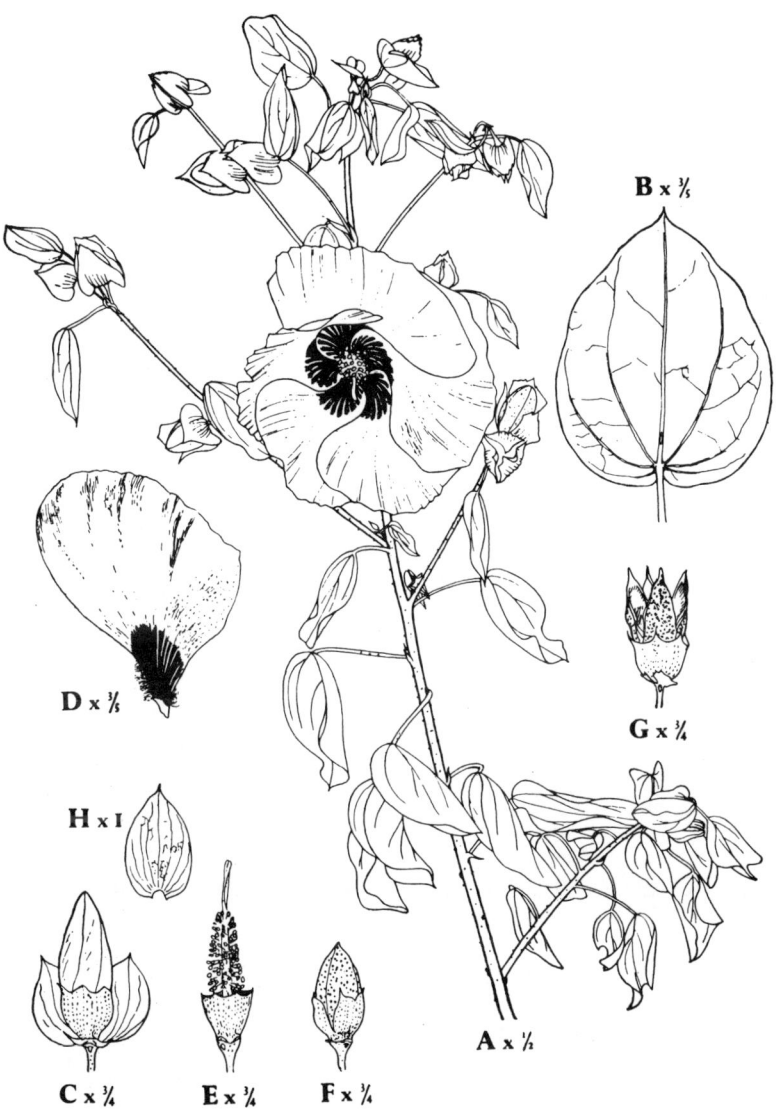

Drawing by J. H. Saunders

Fig. 16 *Gossypium sturtianum.* A, stem and fruiting branches; B, lower surface of flattened leaf, showing nectary on main vein midlobe; C, flower just before opening; D, petal showing spot with rays of pigment; E, pistil and androecium; F, mature capsule; G, dehiscent capsule; H, single entire bracteole. (Reprinted from Saunders, 1961, pl. 3; given there as *Gossypium sturtii.*)

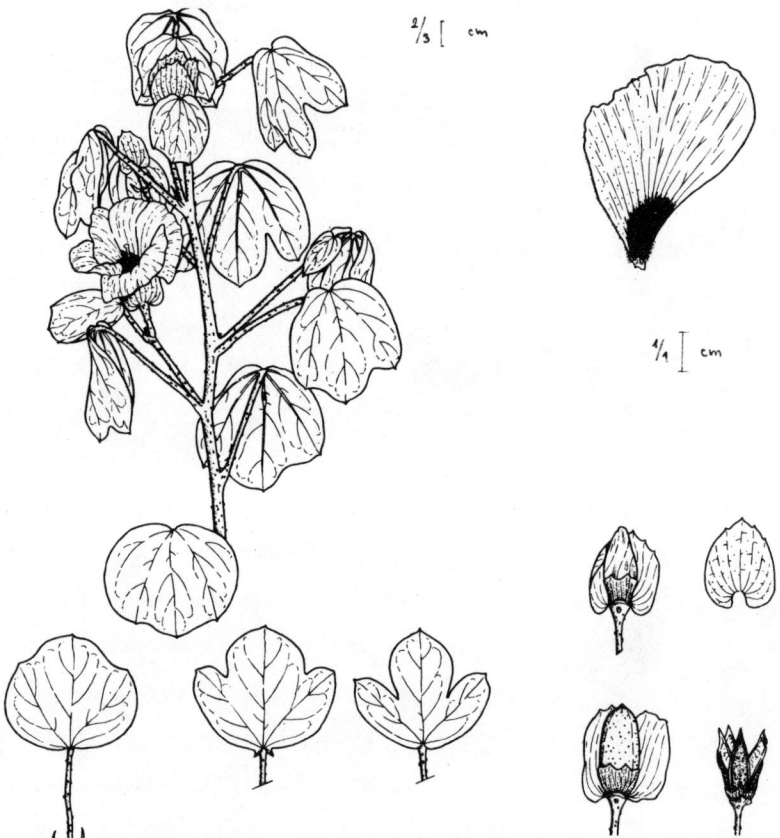

Fig. 17. *Gossypium sturtianum* var. *nandewarense*. (Reprinted from Valíček, 1974, pls. 44, 45.)

1961, pl. 3 (reprinted without attribution by Ter-Avanesyan, 1973, pl. 10); Valíček, 1974, pl. 43.

Distribution: central Australia.

2. *Gossypium sturtianum* var. *nandewarense* (Derera) Fryxell, Bot. Gaz. 125:108. 1964. [Fig. 17]

Gossypium nandewarense Derera, Empire Cotton Growing Rev. 41:14. 1964.

Illustration: Valíček, 1974, pls. 44, 45.

Distribution: southeastern Australia.

3. *Gossypium robinsonii* F. von Mueller, Fragm. 9:126. 1875. [Fig. 18]

Gossypium walchottianum Todaro, Relaz. 119. 1877.

Hibiscus robinsonii (F. von Mueller) Kuntze, Rev. Gen. Pl. 1:69. 1891.

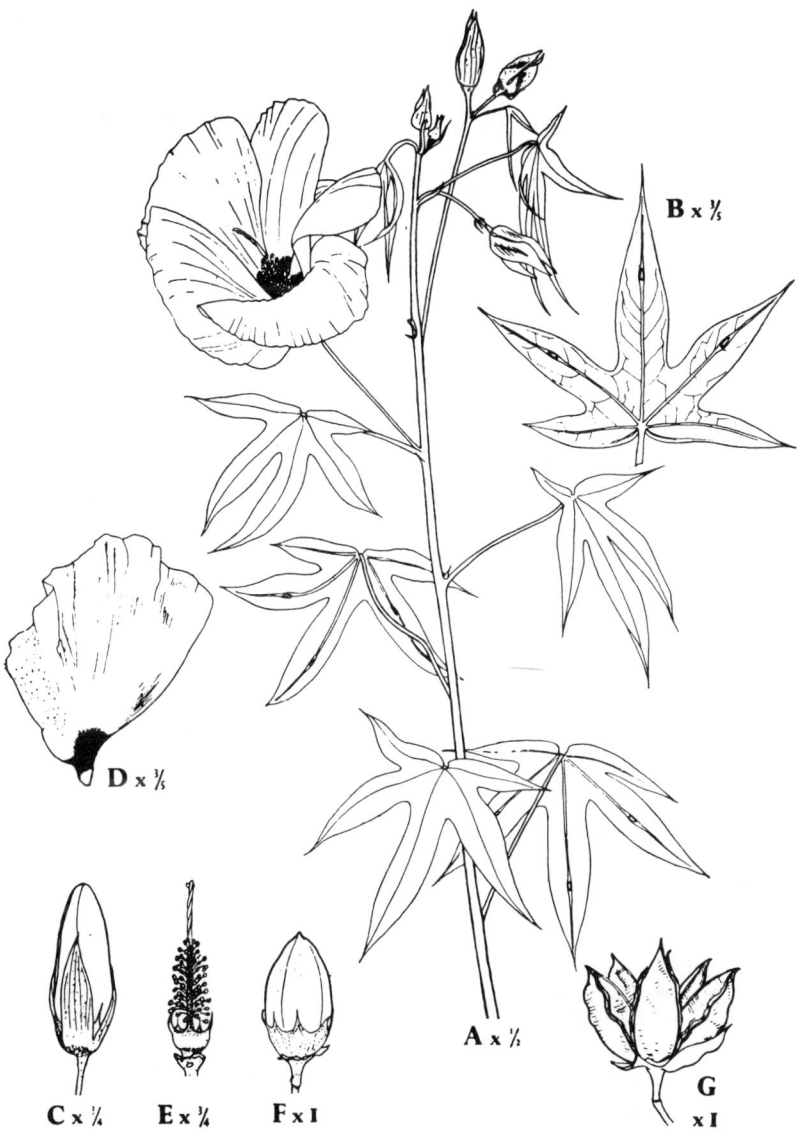

Drawing by J. H. Saunders

Fig. 18. *Gossypium robinsonii.* A, stem and fruiting branches; B, lower surface of flattened leaf, showing three nectaries on main veins of three lobes; C,
flower just before opening, with bracteole showing relative size and shape; D,
petal showing extent of spot; E, pistil and androecium; F, mature boll; G,
dehiscent boll. (Reprinted from Saunders, 1961, pl. 4.)

Cienfuegosia robinsonii (F. von Mueller) Hochreutiner, Ann. Cons. Jard. Bot. Genève 6:57. 1902.
Notoxylinon robinsonii (F. von Mueller), Lewton, J. Wash. Acad. Sci. 5:307. 1915.
Gossypium sturtii ssp. *robinsonii* (F. von Mueller) Roberty, Candollea 13:24. 1950.
Illustrations: Saunders, 1961, pl. 4 (reprinted without attribution by Ter-Avanesyan, 1973, pl. 11); Valíček, 1974, pl. 46.
Distribution: Western Australia.

Section II. *Grandicalyx* Fryxell
Grandicalyx stat. nov. (= subsection *Grandicalyx* Fryxell, Austral. J. Bot. 13:85. 1965.)
Type species: *Gossypium costulatum* Todaro.

4. *Gossypium costulatum* Todaro, Relaz. 109. 1877.
Fugosia latifolia Bentham, Fl. Austral. 1:221. 1863.
Hibiscus latifolius (Bentham) Kuntze, Rev. Gen. Pl. 1:69. 1891.
Cienfuegosia latifolia (Bentham) Hochreutiner, Ann. Cons. Jard. Bot. Genève 6:57. 1902.
Notoxylinon latifolium (Bentham) Lewton, J. Wash. Acad. Sci. 5:307. 1915.
Distribution: northwestern Australia.

5. *Gossypium pulchellum* (C. A. Gardner) Fryxell, Austral. J. Bot. 13:92. 1965.
Fugosia pulchella C. A. Gardner, Forest Dept. Bull. W. Austral. No. 32:63. 1923.
Cienfuegosia pulchella (C. A. Gardner) C. A. Gardner, Enum. Plants Austral. Occid. 79. 1931.
Illustration: Fryxell, 1965a, pl. 2.
Distribution: northwestern Australia.

6. *Gossypium populifolium* (Bentham) F. von Mueller ex Todaro, Relaz. 107. 1877.
Fugosia populifolia Bentham, Fl. Austral. 1:221. 1863.
Hibiscus populifolius (Bentham) Kuntze, Rev. Gen. Pl. 1:69. 1891 (non *H. populifolius* Salisbury).
Cienfuegosia populifolia (Bentham) Lewton, J. Wash. Acad. Sci. 5:306. 1915.
Distribution: northwestern Australia.

7. *Gossypium pilosum* Fryxell, Austral. J. Bot. 22:183. 1974.
Illustration: Fryxell, Austral. J. Bot. 22: figs. 1f–1h. 1974.
Distribution: northwestern Australia.

8. *Gossypium cunninghamii* Todaro, Relaz. 110. 1877.
Fugosia punctata Cunningham ex Bentham, Fl. Austral. 1:220. 1863. (non *F. punctata* Turcz.).

Cienfuegosia benthamii Hochreutiner, Ann. Cons. Jard. Bot. Genève 6:55. 1902.

Notoxylinon punctatum (Cunningham ex Bentham) Lewton, J. Wash. Acad. Sci. 5:307. 1915.

Cienfuegosia punctata (Cunningham ex Bentham) Domin, Biblioth. Bot. 89:964. 1928.

Distribution: northernmost Australia.

Section III. *Hibiscoidea* Todaro
Hibiscoidea Todaro, Relaz. 107. 1877.

Type species: *Gossypium australe* F. von Mueller.

9. *Gossypium australe* F. von Mueller, Fragm. 1:46. 1858.

Fugosia australis (F. von Mueller) Bentham, Fl. Austral. 1:220. 1863.

Cienfuegosia australis (F. von Mueller) K. Schumann in Engler & Prantl, Nat. Plf.-Fam. 3(6):51. 1890.

Hibiscus australis (F. von Mueller) Kuntze, Rev. Gen. Pl. 1:69. 1891.

Notoxylinon australe (F. von Mueller) Lewton, J. Wash. Acad. Sci. 5:307. 1915.

Illustrations: Saunders, 1961, pl. 5 (reprinted without attribution by Ter-Avanesyan, 1973, pl. 12); Valíček, 1974, pl. 47.

Distribution: central Australia.

10. *Gossypium nelsonii* Fryxell, Austral. J. Bot. 22:184. 1974.

Distribution: central Australia.

11. *Gossypium bickii* Prokhanov, Bot. Zhurn. 32:65. 1947.

Fugosia pedata Bailey, Queensl. Agric. J. 25:286. 1910.

Notoxylinon pedatum (Bailey) Lewton, J. Wash. Acad. Sci. 5:307. 1915.

Cienfuegosia pedata (Bailey) Domin, Biblioth. Bot. 89:964. 1928.

Illustration: Valíček, 1974, pl. 48.

Distribution: central Australia.

12. *Gossypium triphyllum* (Harvey) Hochreutiner, Bull. Herb. Boiss. 2:1004. 1902. [Fig. 19]

Fugosia triphylla Harvey ex Harvey & Sonder, Fl. Cap. 2:588. 1862.

Cienfuegosia triphylla (Harvey) Schumann, Bot. Jahrb. Syst. 10:47. 1889.

Gossypium anomalum ssp. *triphyllum* (Harvey) Roberty, Candollea 13:26. 1950.

Illustrations: Saunders, 1961, pl. 2 (reprinted without attribution by Ter-Avanesyan, 1973, pl. 9); Valíček, 1974, pl. 33.

Subgenus *Houzingenia* Fryxell
Houzingenia Fryxell, Taxon 18:587. 1969.

Type species: *Gossypium trilobum* (DC.) Skovsted.

Section IV. *Houzingenia*

Subsection IVa. *Houzingenia*

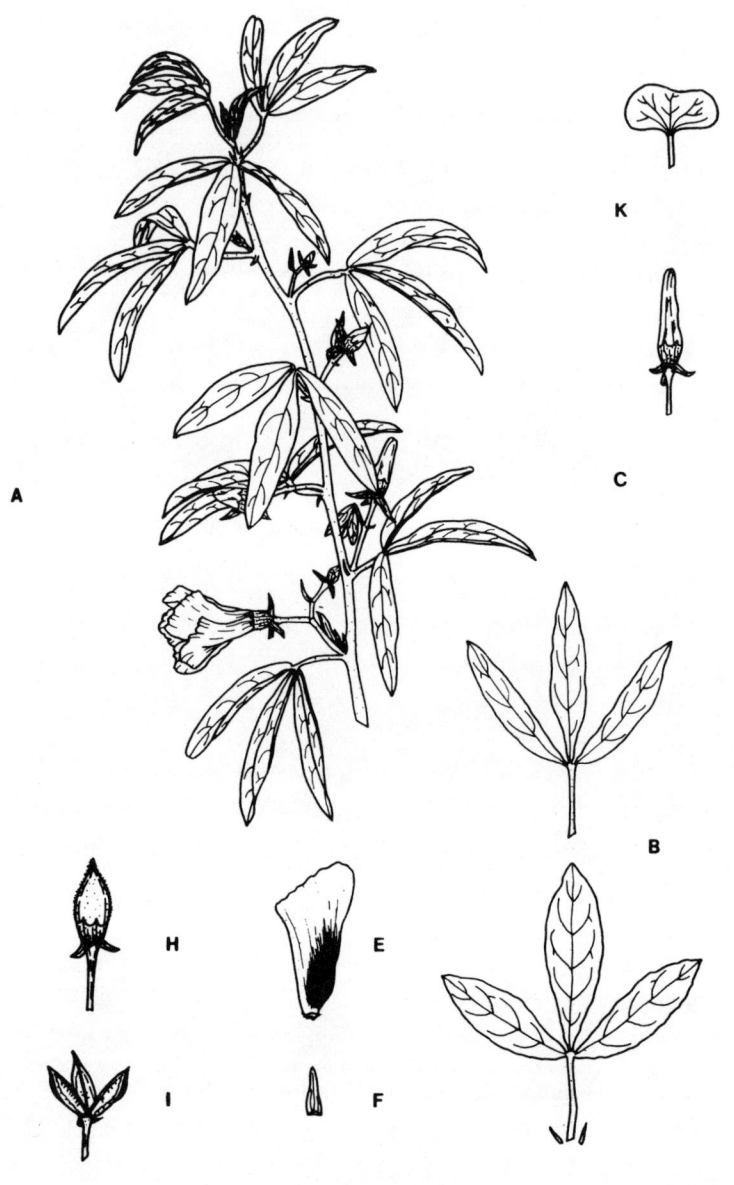

Fig. 19. *Gossypium triphyllum.* A, flowering branch; B, leaves; C, flower bud;
E, petal; F, involucellar bract; H, young fruit; I, mature fruit; K, cotyledon.
(Reprinted from Valíček, 1974, pl. 33.)

Fig. 20. *Gossypium trilobum*. (Reprinted from Wilson, Lee, and Bridge, 1968, fig. 1.)

13. *Gossypium trilobum* (Mociño & Sessé ex DC.) Skovsted, J. Genet. 31:288. 1935 (pro parte) emend. Kearney, Amer. J. Bot. 24:299. 1937. [Fig. 20]

Ingenhouzia triloba Mociño & Sessé ex DC., Prodr. 1:474. 1824.

Hibiscus ingenhousii Kuntze, Rev. Gen. Pl. 1:69. 1891 (pro parte).

Gossypium lanceiforme Miers ex Britten, J. Bot. 31:331. 1893.

Thurberia triloba (Mociño & Sessé ex DC.) Tidestrom ex Dayton, Proc. Biol. Soc.Wash. 40:120. 1927 (pro parte).

Illustrations: Hutchinson 1947*b*, pls. 3*a*, 3*b*; Fryxell and Parks, 1967, figs. 1–2; Wilson, Lee, and Bridge, 1968, fig. 1; Valíček, 1974, pl. 65.

Distribution: western Mexico, Sinaloa to Morelos.

14. *Gossypium thurberi* Todaro, Relaz. 120. 1877. [Fig. 21]
Thurberia thespesioides A. Gray, Mem. Amer. Acad. 5:308. 1855.
Thespesia thurberi Alefeld, Bot. Zeit. 19:301. 1861.
Hibiscus ingenhousii Kuntze, Rev. Gen. Pl. 1:69. 1891 (pro parte).
Thurberia triloba Tidestrom ex Dayton, Proc. Biol. Soc. Wash. 40:120. 1927
(pro parte).
Gossypium trilobum (Mociño & Sessé ex DC.) Skovsted, J. Genet. 31:288.
1935 (pro parte).

Illustrations: Torrey, 1859, pl. 6; Hutchinson, 1947*b*, pl. 1; Mauer, 1954, pl. 90; Saunders, 1961, pl. 6 (reprinted without attribution by Ter-Avanesyan, 1973, pl. 14); Valíček, 1974, pl. 55.

Distribution: Arizona and northern Mexico (Sonora and Chihuahua).

Subsection IVb. *Integrifolia* (Todaro) Todaro
Integrifolia (Todaro) Todaro, Relaz. 188. 1877.

Type species: *Gossypium klotzschianum* Andersson.

15. *Gossypium klotzschianum* Andersson, Kongl. Vetensk. Acad.
Handl. 1853:228. 1855.

Illustrations: Hutchinson, 1947*b*, pl. 2 (reprinted without attribution by Mauer, 1954, pl. 86); Mauer, 1954, pl. 89; Saunders, 1961, pl. 9*i* (reprinted without attribution by Ter-Avanesyan, 1973, pl. 17); Valíček, 1974, pl. 59.

Distribution: Galápagos Islands.

16. *Gossypium davidsonii* Kellogg, Proc. Calif. Acad. Sci. 5:82. 1873.
Gossypium klotzschianum var. *davidsonii* (Kellogg) Hutchinson, Evol. Gossyp.
22. 1947.

Illustrations: Watt, 1907, pl. 3; Hutchinson, 1947*b*, pl. 2 (reprinted without attribution by Mauer, 1954, pl. 86); Mauer, 1954, pl. 87; Saunders, 1961, pl. 9, excl. *i* (reprinted without attribution by Ter-Avanesyan, 1973, pl. 17); Valíček, 1974, pl. 58.

Distribution: Baja California.

Subsection IVc. *Caducibracteolata* Mauer
Caducibracteolata Mauer, Trudy Sredne-Aziatsk. Gosud. Univ. Lenina, n.s.,
18:21. 1950.

Type species: *Gossypium harknessii* Brandegee.

17. *Gossypium harknessii* Brandegee, Proc. Calif. Acad. Sci., ser. 2,
2:136. 1889.
Ingenhouzia harknessii (Brandegee) Rose ex Tyler, U.S.D.A. Bur. Plant In-
dus. Bull. 131:54. 1908.

Fig. 21. *Gossypium thurberi*. (Reprinted from Torrey, 1859, pl. 6; given there as *Thurberia thespesioides*.)

Gossypium californicum Mauer, Trudy Sredne-Aziatsk. Gosud. Univ. Lenina, n.s., 7:21. 1950.

Illustrations: Hutchinson, 1947*b*, pl. 1; Mauer, 1954, pls. 97–98; Saunders, 1961, pl. 8 (reprinted without attribution by Ter-Avanesyan, 1973, pl. 16); Valíček, 1974, pl. 57.

Distribution: Baja California.

18. *Gossypium armourianum* Kearney, J. Wash. Acad. Sci. 23:558. 1933.

Gossypium harknessii ssp. *armourianum* (Kearney) Roberty, Candollea 13:25. 1950.

Illustrations: Hutchinson, 1947*b*, pl. 1; Mauer, 1954, pl. 96; Saunders, 1961, pl. 7 (reprinted without attribution by Ter-Avanesyan, 1973, pl. 15); Valíček, 1974, pl. 56.

Distribution: San Marcos Island and on peninsular Baja California.

19. *Gossypium turneri* Fryxell, Madroño 25:155, fig. 1. 1978.

Distribution: near Guaymas, Sonora, Mexico.

Section V. *Erioxylum* (Rose & Standley) Prokhanov
Erioxylum (Rose & Standley) Prokhanov, Bot. Zhurn. 32:71. 1947.

Type species: *Gossypium aridum* (Rose & Standley) Skovsted.

Subsection Va. *Erioxylum* (Rose & Standley) Fryxell
Erioxylum (Rose & Standley) Fryxell, Taxon 18:588. 1969.

20. *Gossypium aridum* (Rose & Standley) Skovsted, Ann. Bot. (London) 47:28. 1933. [Fig. 22]
Erioxylum aridum Rose & Standley, Contr. U.S. Natl. Herb. 13:307. 1911.
Gossypium rosei Prokhanov, Bot. Zhurn. 32:72. 1947.
Gossypium aridum var. *palmeri* (Rose) Mauer, Proiskhozhd. i Sist. Khlopchat. [Orig. and Syst. of Cott.] 265. 1954.

Illustrations: Hutchinson, 1947*b*, pl. 1; Saunders, 1961, pl. 10 (reprinted without attribution by Ter-Avanesyan, 1973, pl. 18); Mauer, 1954, pls. 92–93; Valíček, 1974, pl. 60 (actually an *aridum* × *lobatum* hybrid).

Distribution: western Mexico (Sinaloa to Oaxaca).

21. *Gossypium laxum* Phillips, Madroño 21:265. 1972.

Illustration: Valíček, 1974, pl. 66.

Distribution: Guerrero, Mexico.

22. *Gossypium lobatum* Gentry, Madroño 13:261. 1956. [Fig. 23]
Illustrations: Gentry, Madroño 13: fig. 1 (reprinted by Saunders, 1961, pl. 13, and without attribution by Ter-Avanesyan, 1973, pl. 21, in part); Valíček, 1974, pl. 64.

Distribution: Michoacán, Mexico (see Fryxell and Koch, 1978).

Subsection Vb. *Selera* (Ulbrich) Fryxell
Selera (Ulbrich) Fryxell, Taxon 18:588. 1969.

Type species: *Gossypium gossypioides* (Ulbrich) Standley.

23. *Gossypium gossypioides* (Ulbrich) Standley, Contr. U.S. Natl. Herb. 23:783. 1923 (non *G. gossypioides* [R. Brown] C. A. Gardner). [Fig. 24]
Selera gossypioides Ulbrich, Verh. Bot. Vereins Prov. Brandenburg 55:51. 1913.

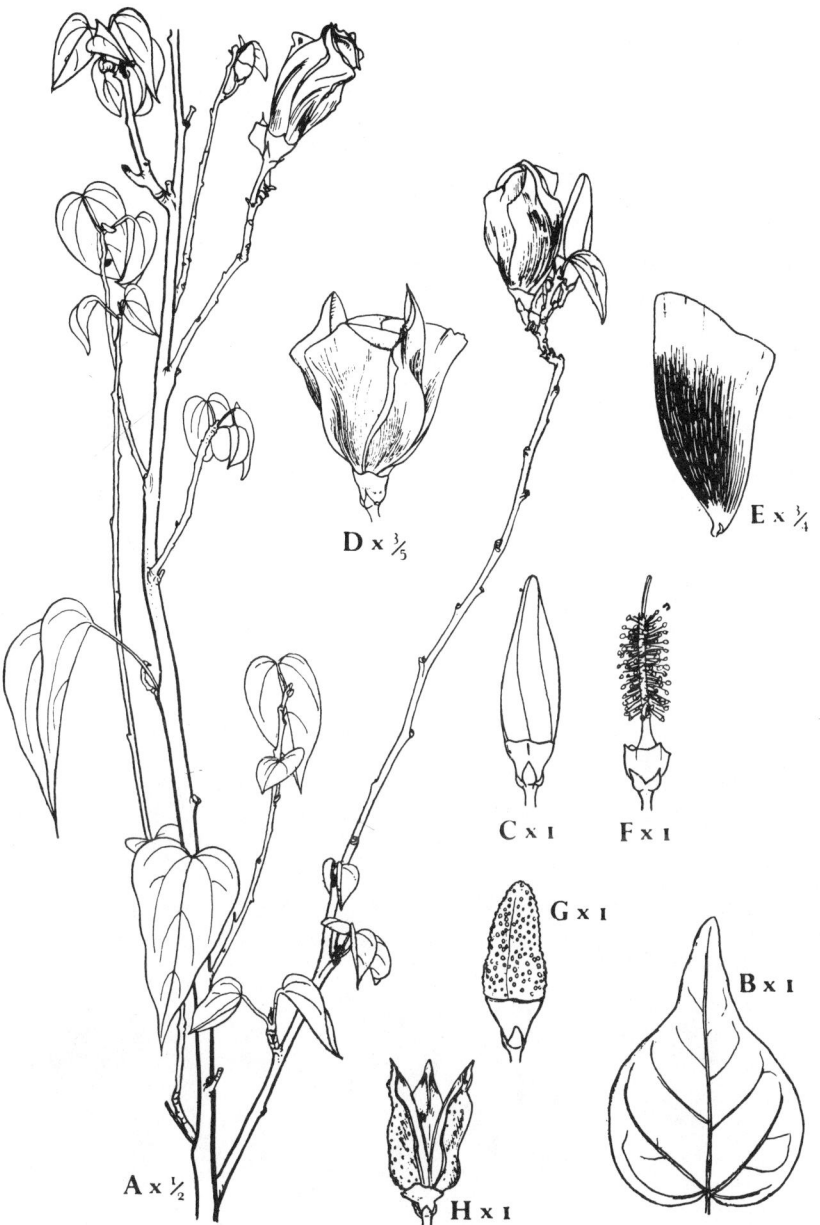

Fig. 22. *Gossypium aridum.* A, stem and fruiting branches; B, lower surface of flattened leaf, showing nectary on main vein; C, flower just before opening (note very small bracteole); D, tubular cup-shaped flower; E, narrow petal with huge spot; F, pistil and androecium; G, mature boll (note prominent glands); H, narrowly dehiscent boll. Note that at the time of flowering the leaves are shedding. (Reprinted from Saunders, 1961, pl. 10.)

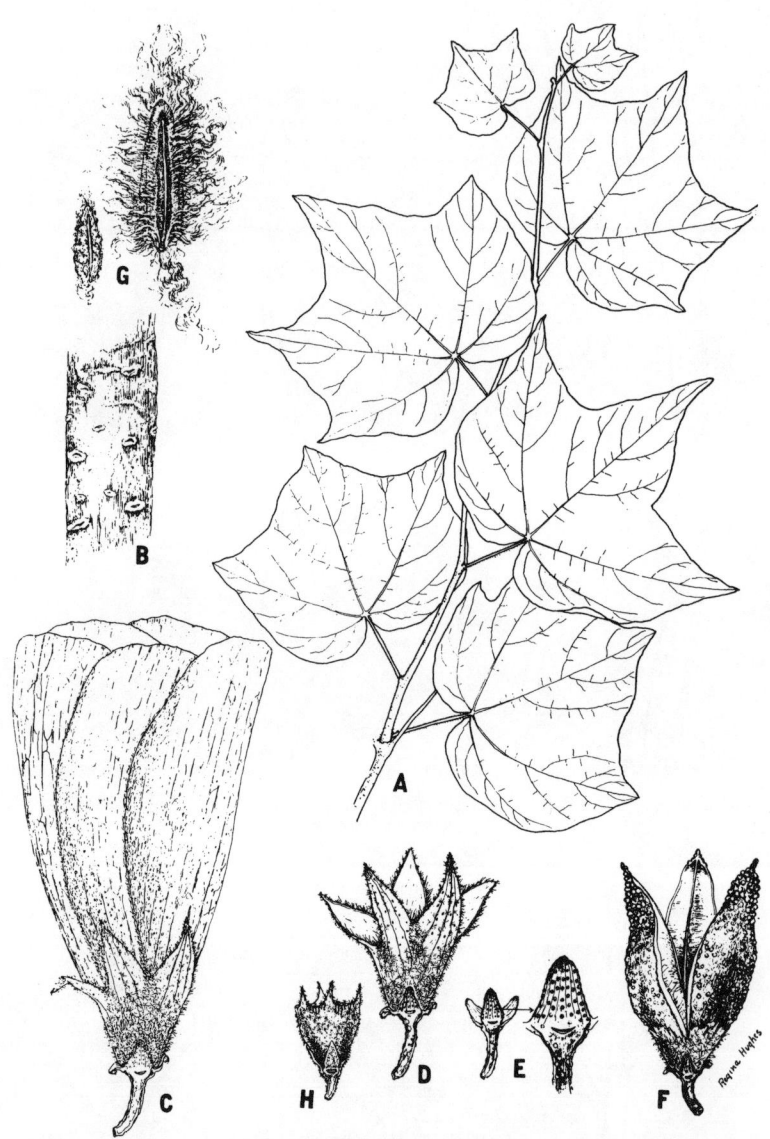

Fig. 23. *Gossypium lobatum.* A, foliage branch (× 1/3); B, section of bark (× 2); C, flower (× 1); D, calyx (×1); E, involucel (× 1 and× 2); F, capsule (× 1); G, seed (× 1); H, *G. aridum* calyx (× 1). (Reprinted from Gentry, Madroño 13: fig. 1. 1956.)

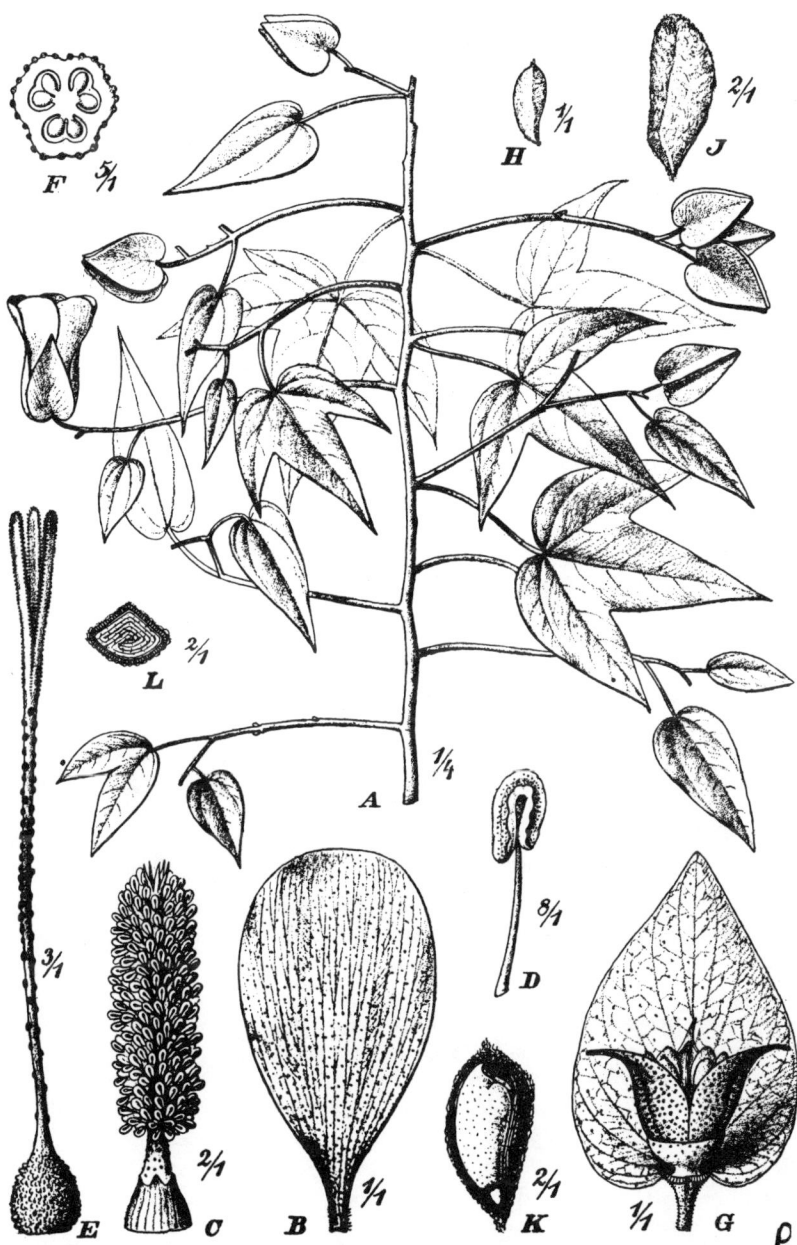

Fig. 24. *Gossypium gossypioides*. A, stem and branches; B, petal; C, pistil and androecium; D, single anther; E, ovary and style; F, cross-section through the ovary; G, open capsule with cup-shaped calyx and obdurate outer calyx (the two foremost blades of the outer calyx are deleted); H and J, seeds; K, longitudinal section of the embryo showing the folded cotyledon; L, cross-section of the embryo. (Reprinted from Ulbrich, Verh. Bot. Vereins Prov. Brandenburg 55:169. 1913; given there as *Selera gossypioides*.)

Gossypium trilobum (DC.) Skovsted emend. Roberty, Candollea 13:30. 1950 (pro parte).

Illustrations: Ulbrich, Verh. Bot. Vereins Prov. Brandenburg 55:169. 1913 (reprinted by Mauer, 1954, pl. 94); Saunders, 1961, pl. 12 (reprinted without attribution by Ter-Avanesyan, 1973, pl. 20); Valíček, 1974, pl. 63.

Distribution: Oaxaca, Mexico.

Subsection Vc. *Austroamericana* Fryxell

Austroamericana Fryxell, Taxon 18:589. 1959.

Type species: *Gossypium raimondii* Ulbrich.

24. *Gossypium raimondii* Ulbrich, Notizbl. Bot. Gart. Berlin-Dahlem 11:548. 1932. [Fig. 25]

Gossypium klotzschianum ssp. *raimondii* (Ulbrich) Roberty, Candollea 13:29. 1950.

Illustrations: Hutchinson, 1947*b*, pl. 2 (reprinted without attribution by Mauer, 1954, pl. 86, fig. 3); Saunders, 1961, pl. 11 (reprinted without attribution by Ter-Avanesyan, 1973, pl. 19); Valíček, 1974, pls. 61–62.

Distribution: Peru.

Subgenus *Gossypium*

Type species: *Gossypium arboreum* L.

Section VI. *Gossypium*

Subsection VIa. *Gossypium*

25. *Gossypium arboreum* L., Sp. Pl. 693. 1753.

Gossypium rubrum Forskål, Fl. Aegypt.-Arab. 125. 1775.

Gossypium indicum Medikus, Bot. Beobacht. Jahr. 1783 197. 1784.

Gossypium rufum Scopoli, Delioc. Insub. 3:70. 1788.

Gossypium nigrum Buchanan-Hamilton, Trans. Linn. Soc. 13:494. 1822.

Gossypium croceum Buchanan-Hamilton, *loc. cit.*

Gossypium obtusifolium Roxburgh ex G. Don, Gen. Hist. 1:487. 1831.

Gossypium nanking Meyen, Verh. Ver. Beförd. Gartenb. Königl. Preuss. Staat. 11:258, t. 3. 1835.

Gossypium speciosum Rafinesque, Sylva Tellur. 18. 1838.

Gossypium puniceum Fenzl ex Jacquin, Eclog. Pl. Rar. 2:7, t. 134. 1844.

Gossypium sanguineum Hasskarl, Cat. Hort. Bogor. 200. 1844.

Gossypium albiflorum Todaro, Giorn. R. Ist. Incoragg. Agric. Arti Manifatture Sicil., ser. 3, 1(2, 3):42. 1863.

Gossypium cernuum Todaro, *op. cit.*, 47.

Gossypium neglectum Todaro, *op. cit.*, 51.

Gossypium royleanum Todaro, *op. cit.*, 57.

Gossypium intermedium Todaro, *op cit.*, 58.

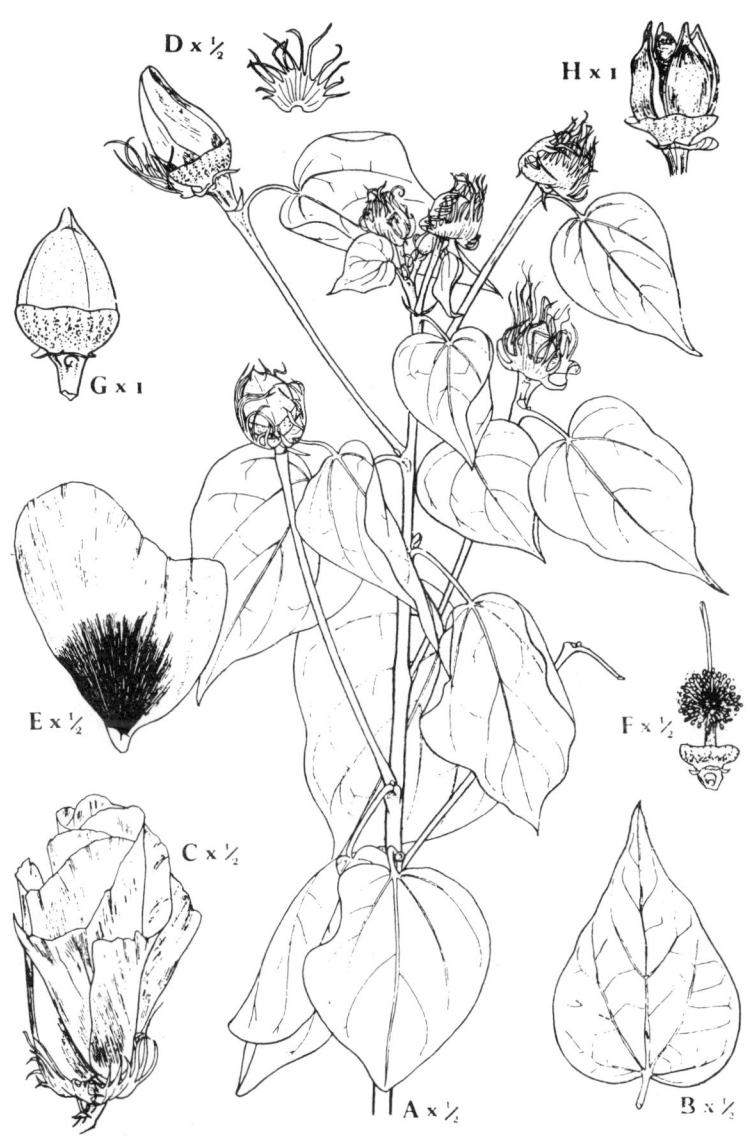

Drawing by J. H. Saunders

Fig. 25. *Gossypium raimondii.* A, stem and fruiting branches; B, lower surface of flattened leaf, showing nectary on main vein; C, open tubular flower; D, flower just before opening and single bracteole; E, petal, showing very large spot; F, pistil and androecium; G, mature capsule; H, dehiscent capsule. (Reprinted from Saunders, 1961, pl. 11.)

Gossypium roxburghii Todaro, *op. cit.*, 61.
Gossypium wightianum Todaro, *op. cit.*, 63.
Hibiscus cernuus (Todaro) Kuntze, Rev. Gen. Pl. 1:68. 1891.
Hibiscus nanking (Meyen) Kuntze, *loc. cit.*
Gossypium soudanense (Watt) Watt, Kew Bull. 5:201. 1926.
Gossypium bani (Watt) Prokhanov, Bot. Zhurn. 32:69. 1947.
Gossypium perrieri (Hochreutiner) Prokhanov, *op. cit.*, 71.
Gossypium wattianum Hu, Fl. China, fam. 153 (Malv.), 65. 1955.

Illustrations: Parlatore, 1866, pl. 1; Todaro, 1878, pls. 1, 2, 3; Gammie, 1905, pls. 1, 6, 7, 8, 9; Watt, 1907, pls. 8, 13, 16; Hutchinson, 1947*b*, pl. 4; Valíček, 1974, pls. 29, 30.

Distribution: India and the Far East, in cultivation (occasionally elsewhere in cultivation).

26. *Gossypium herbaceum* L., Sp. Pl. 693. 1753.
Gossypium frutescens Lasteyrie, Du Cottonier 435, t. 1. 1808.
Gossypium album Buchanan-Hamilton, Trans. Linn. Soc. 13:494. 1822.
Gossypium africanum (Watt) Watt, Kew Bull. 5:205. 1926.
Gossypium transvaalense Watt, *op. cit.*, 207.
Gossypium abyssinicum Watt, *op. cit.*, 208.
Gossypium zaitzevii Prokhanov, Bot. Zhurn. 32:70. 1947.

Illustrations: Parlatore, 1866, pl. 2; Todaro, 1878, pl. 4; Gammie, 1905, pls. 3, 4; Watt, 1907, pls. 22, 25; Hutchinson, 1947*b*, pl. 5; Mauer, 1954, pls. 53, 59, 60; Saunders, 1961, pl. 19 (reprinted without attribution by Ter-Avanesyan, 1973, pl. 27); Valíček, 1974, pls. 26, 27.

Distribution: cultivated in Middle East and in parts of Africa (occasionally elsewhere in cultivation); an apparently wild form indigenous in South Africa.

Subsection VIb. *Anomala* Todaro
Anomala Todaro, Relaz. 120. 1877.

Type species: *Gossypium anomalum* Wawra ex Wawra & Peyritch.

27. *Gossypium anomalum* Wawra ex Wawra & Peyritch, Sitzungsber. Kaiserl. Akad. Wiss. Math.-Naturwiss. Cl. 38:561. 1860 (non Watt).
Hibiscus anomalus (Wawra & Peyritch) Kuntze, Rev. Gen. Pl. 1:68. 1891.
Cienfuegosia pentaphylla Schumann, Bot. Jahrb. Syst. 10:48. 1889.
Gossypium senarense Fenzl ex Wawra & Peyritsch, *loc. cit.*
Gossypium microcarpum Welwitsch ex Gürke, Bot. Jahrb. Syst. 19, Beibl. 48(2):2. 1894 (pro syn.).
Gossypium herbaceum var. *steudneri* Schweinfurth ex Gürke, *loc. cit.* (pro syn.).
Gossypium anomalum ssp. *steudneri* (Schweinfurth ex Gürke) Roberty, Candollea 13:27. 1950.

Illustrations: Todaro, 1878, pl. 5 in part; Hutchinson, 1947*b*, pl. 4; Knight, 1949, pl. 1; Mauer, 1954, pl. 82; Saunders, 1961, pl. 1 (reprinted without attribution by Ter-Avanesyan, 1973, pl. 8); Valíček, 1974, pl. 32.

Distribution: disjunct in Africa: (1) in Angola and Namibia and (2) from Niger to the Sudan.

28. *Gossypium capitis-viridis* Mauer, Trudy Sredne-Aziatsk. Gosud. Univ. Lenina, n.s., 18:19. 1950.
Gossypium barbosanum Phillips & Clement, Bot. Mus. Leafl. 20:214. 1963.

Illustration: Valíček, 1974, pl. 34.

Distribution: Cape Verde Islands.

Section VII. *Pseudopambak* Prokhanov
Pseudopambak Prokhanov, Bot. Zhurn. 32:65. 1947.

Type species: *Gossypium stocksii* Masters in Hooker.

Subsection VIIa. *Pseudopambak*

29. *Gossypium stocksii* Masters in Hooker, Fl. Brit. Ind. 1:346. 1874. [Fig. 26]

Illustrations: Watt, 1907, pl. 6; Hutchinson, 1947*b*, pl. 4; Mauer, 1954, pl. 84; Saunders, 1961, pl. 14 (reprinted without attribution by Ter-Avanesyan, 1973, pl. 22); Valíček, 1974, pl. 36.

Distribution: Sind (Pakistan), Oman (Arabia), and Somalia.

30. *Gossypium somalense* (Gürke) J. B. Hutchinson, Evol. Gossyp. 31. 1947.
Cienfuegosia somalensis Gürke, Bot. Jahrb. Syst. 33:380. 1904.
Cienfuegosia ellenbeckii Gürke, *op. cit.*, 381.
Cienfuegosia bricchettii Ulbrich, Bot. Jahrb. Syst. 48:378. 1912.
Gossypium benadirense Mattei, Boll. Stud. Inform. Reale Giardino Colon. 2:223. 1916.
Gossypium paolii Mattei, *loc. cit.*
Gossypium ellenbeckii (Gürke) Mauer, Trudy Sredne-Aziatsk. Gosud. Univ. Lenina, n.s., 18:19. 1950.

Illustrations: Knight, 1949, pl. 2 (reprinted by Mauer, 1954, pl. 80); Saunders, 1961, pl. 15 (reprinted without attribution by Ter-Avanesyan, 1973, pl. 23); Valíček, 1974, pl. 37.

Distribution: Somalia, Kenya, and the Sudan.

31. *Gossypium incanum* (Schwartz) Hillcoat, Empire Cotton Growing Rev. 36:165. 1959.
Cienfuegosia incana Schwartz, Mitt. Inst. Allg. Bot. Hamburg 10:165. 1935.

Illustrations: Saunders, 1961, pl. 17 (reprinted without attribution by Ter-Avanesyan, 1973, pl. 25); Valíček, 1974, pl. 39.

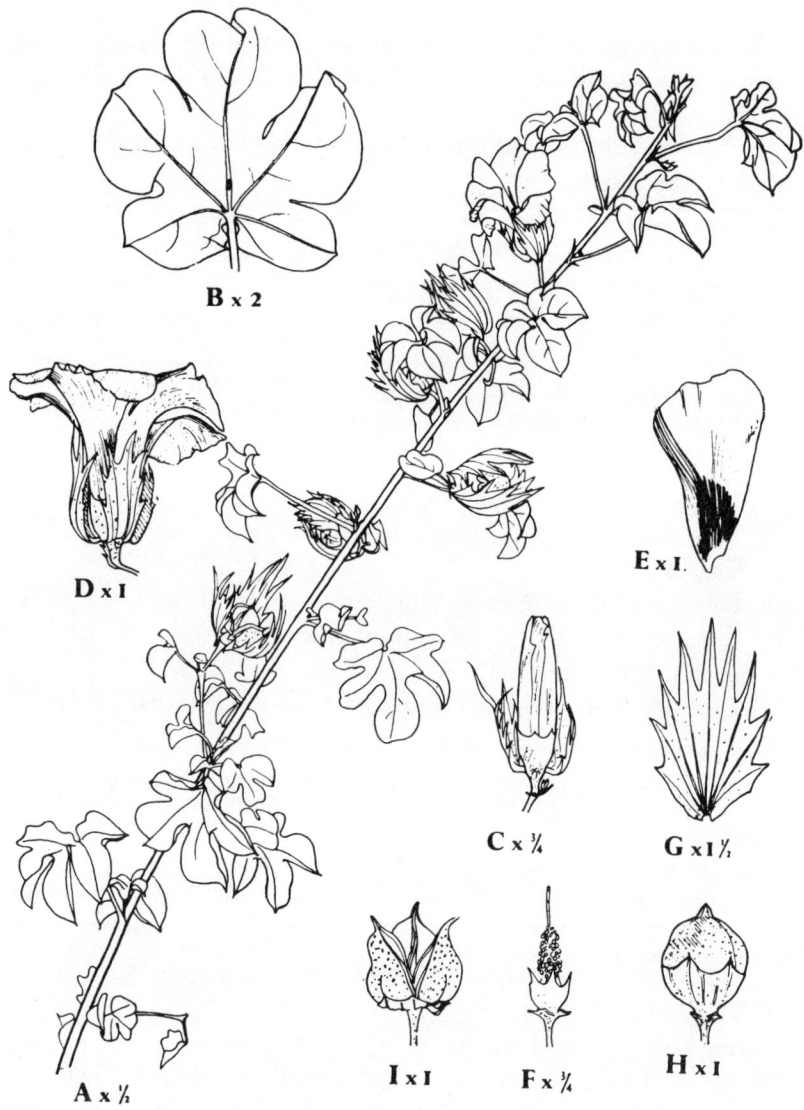

Drawing by J. H. Saunders

Fig. 26. *Gossypium stocksii.* A, stem and fruiting branches; B, lower surface of flattened leaf, showing nectary on main vein midlobe; C, flower just before opening; D, open campanulate flower; E, petal, showing extent of spot; F, pistil and androecium; G, nine-toothed bracteole; H, mature capsule; I, dehiscent capsule. (Reprinted from Saunders, 1961, pl. 14.)

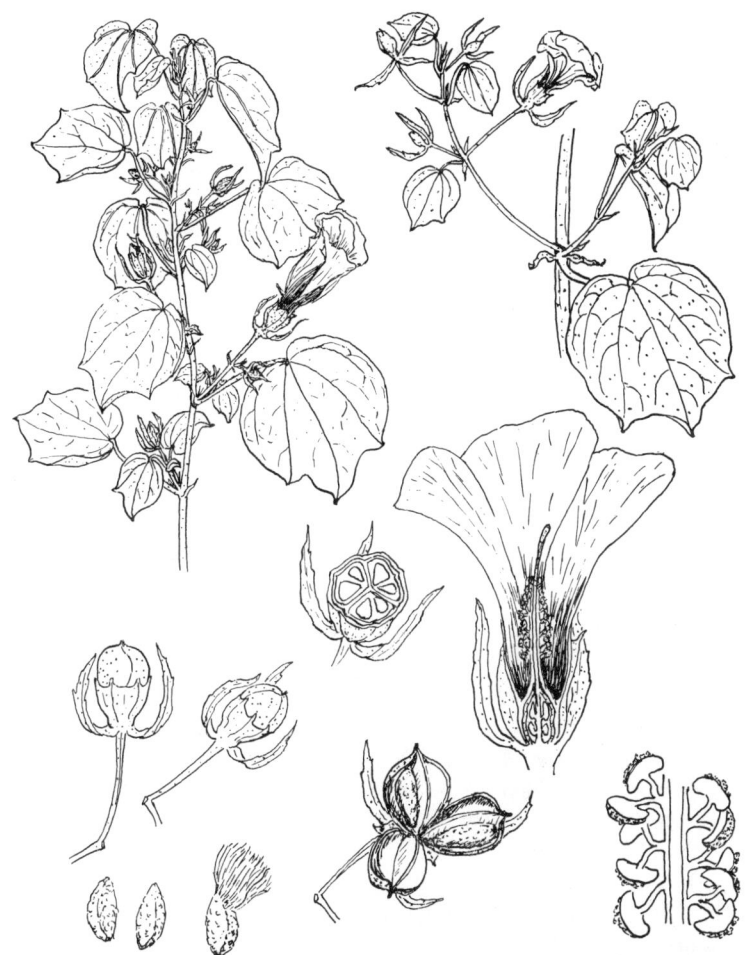

Fig. 27. *Gossypium areysianum.* (Reprinted from Douwes, 1953, fig. 1.)

Distribution: southern Arabia (Hearn, 1968).

32. *Gossypium areysianum* Deflers, Esq. Geogr. Bot. 49. 1895. [Fig. 27]

Fugosia areysiana (Deflers) Deflers, Bull. Soc. Bot. France 42:299. 1895.

Gossypium anomalum ssp. *areysianum* (Deflers) Roberty, Candollea 13:27. 1950.

Illustrations: Douwes, 1953, fig. 1 (reprinted by Mauer, 1954, pl. 85); Saunders, 1961, pl. 16 (reprinted without attribution by Ter-Avanesyan, 1973, pl. 24); Valíček, 1974, pl. 38.

Distribution: southern Arabia (Hearn, 1968).

Subsection VIIb. *Longiloba* Fryxell
Longiloba Fryxell, Taxon 18:590. 1969.

Type species: *Gossypium longicalyx* Hutchinson & Lee.

33. *Gossypium longicalyx* Hutchinson & Lee, Kew Bull. 1958:221.
1958. [Fig. 28]
Illustrations: Saunders, 1961, pl. 18 (reprinted without attribution by
Ter-Avanesyan, 1973, pl. 26); Valíček, 1974, pl. 4.
Distribution: East Africa.

Subgenus *Karpas* Rafinesque
Karpas Rafinesque, Sylva Tellur. 14. 1838.

Type species: *Gossypium guyanense* Rafinesque.

34. *Gossypium tomentosum* Nuttall ex Seemann, Fl. Vit. 22. 1865.
Gossypium sandvicense Parlatore, Sp. Cot. 37, t. 6. 1866.
Hibiscus tomentosus (Nuttall ex Seemann) Kuntze, Rev. Gen. Pl. 1:68. 1891.
Gossypium tomentosum var. *parvifolium* Nuttall ex Watt, Wild and Cult. Cot-
ton Plants 71. 1907.
Gossypium hirsutum f. *tomentosum* (Nuttall ex. Seemann) Roberty, Candollea
13:73. 1950.

Illustrations: Parlatore, 1866, pl. 6*b*; Watt, 1907, pl. 5; Mauer, 1954,
pl. 126; Valíček, 1974, pl. 75.
Distribution: Hawaiian Islands.

35. *Gossypium lanceolatum* Todaro, Relaz. 185. 1877.
Gossypium fruticulosum Todaro, *op. cit.*, 187.
Gossypium janiphaefolium Bello, Anal. Soc. Esp. Hist. Nat. 10:242. 1881.
Gossypium palmeri Watt, Wild and Cult. Cotton Plants 204, pl. 34. 1907.
Illustration: Watt, 1907, pl. 34.
Distribution: Mexico, principally in dooryard cultivation.

36. *Gossypium hirsutum* L., Sp. Pl. 975. 1763.
Gossypium religiosum L., Syst. Nat. 462. 1767.
Gossypium latifolium Murray, Nov. Comment. Soc. Reg. Sc. Götting. 7:22, t.
1. 1776.
Gossypium micranthum Cavanilles, Diss. 6:311, t. 193. 1788.
Gossypium siamense Tussac, Fl. Antill. 2:68. 1818.
Gossypium punctatum Schumacher & Thonning in Schumacher, Beskr. Guin.
Plant. 309. 1827.
Gossypium prostratum Schumacher & Thonning in Schumacher, *op. cit.*, 311.
Gossypium paniculatum Blanco, Fl. Filip. 539. 1837.
Gossypium jamaicense Macfadyen, Fl. Jam. 1:73. 1837.
Gossypium oligospermum Macfadyen, *op. cit.*, 74.
Gossypium asiaticum Rafinesque, Sylva Tellur. 16. 1838.
Gossypium pallens Rafinesque, *loc. cit.*

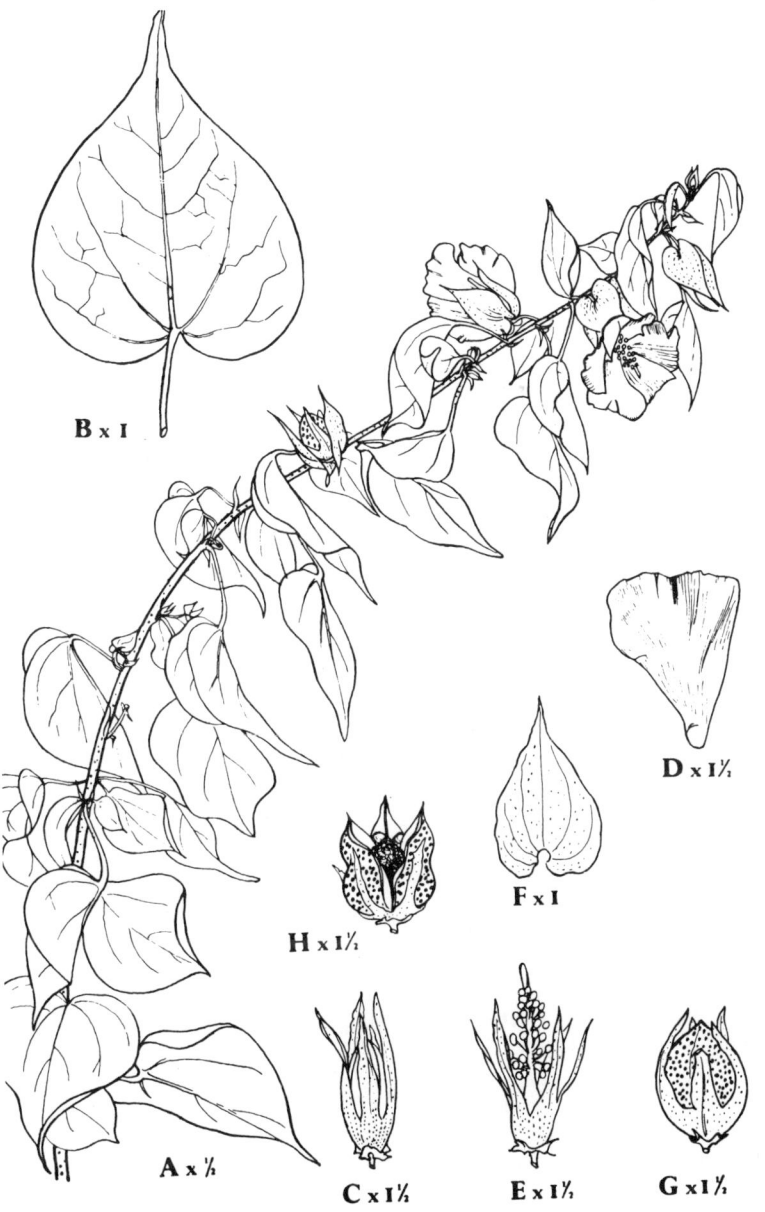

B x I

D x 1½

F x I

H x 1½

A x ½

C x 1½   E x 1½   G x 1½

Drawing by J. H. Saunders

Fig. 28. *Gossypium longicalyx.* A, stem and fruiting branches; B, flattened leaf; C, flower just before opening (note relative length of calyx teeth); D, petal, without spot; E, pistil and androecium; F, entire bracteole; G, mature capsule; H, dehiscent capsule. (Reprinted from Saunders, 1961, pl. 18.)

Gossypium convexum Rafinesque, *loc. cit.*

Gossypium divaricatum Rafinesque, *op. cit.*, 17.

Gossypium caespitosum Todaro, Giorn. R. Ist. Incoragg. Agric. Arti Manifatture Sicil. 3, 1:80. 1863.

Gossypium taitense Parlatore, Sp. Cot. 39, t. 6*a*, 1866.

Gossypium mexicanum Todaro, Ind. Sem. Hort. Bot. Panorm. 1867:20, 31. 1868.

Hibiscus fruticulosus (Todaro) Kuntze, Rev. Gen. Pl. 1:68. 1891.

Hibiscus religiosus (L.) Kuntze, *loc. cit.*

Gossypium schottii Watt, Wild and Cult. Cotton Plants 206. 1907.

Gossypium nicaraguense Ramirez-Goyena, Fl. Nicar. 1:195. 1909.

Gossypium volubile Ramirez-Goyena, *loc. cit.*

Gossypium hopi Lewton, Smithson. Misc. Coll. 60(6):9. 1912.

Gossypium nervosum Watt, Kew Bull. 8:324. 1927.

Gossypium birkinshawii Watt, Kew Bull. 8:330. 1927.

Gossypium harrissii Watt, Kew Bull. 8:331. 1927.

Gossypium marie-galante Watt, Kew Bull. 8:344. 1927.

Gossypium sericatum Prokhanov, Bot. Zhurn. 32:73. 1947.

Illustrations: Parlatore, 1866, pls. 5, 6*a*; Todaro, 1878, pl. 6; Gammie, 1905, pl. 5; Watt, 1907, pls. 28, 30, 43; Hutchinson, 1947*b*, pls. 6, 7; Valíček, 1974, pl. 67.

Distribution: indigenous to Middle America and the Caribbean and in certain Pacific Islands (Socorro, the Marquesas, Samoa, etc.); now cosmopolitan in cultivation.

37. *Gossypium barbadense* L., Sp. Pl. 693. 1753.

Gossypium vitifolium Lamarck, Encycl. Meth. Bot. 2:135. 1786.

Gossypium peruvianum Cavanilles, Diss. 6:313, t. 168. 1788.

Gossypium lapideum Tussac, Fl. Antill. 2:67. 1818.

Gossypium acuminatum Roxburgh ex Don, Gen. Syst. 1:487. 1831.

Gossypium suffruticosum Bertoloni, Nov. Comm. Acad. Sci. Bonon. 2:216, t. 9. 1836.

Gossypium brasiliense Macfadyen, Fl. Jamaic. 1:72. 1837.

Gossypium perenne Blanco, Fl. Filip. 537. 1837.

Gossypium rohrianum Rafinesque, Sylva Tellur. 19. 1838.

Gossypium guyanense Rafinesque, *op. cit.*, 16.

Gossypium vaupellii Graham, Cat. Pl. Bomb. 15. 1839.

Gossypium maritimum Todaro, Giorn. R. Ist. Incoragg. Agric. Arti Manifatture Sicil. 3, 1:100. 1863.

Gossypium multiglandulosum Phillipi, Anal. Mus. Nac. Chile 2:10. 1891.

Hibiscus barbadensis (L.) Kuntze, Rev. Gen. Pl. 1:67. 1891.

Gossypium quinacre Cook & Hubbard, J. Wash. Acad. Sci. 16:548. 1926.

Gossypium pedatum Watt, Kew Bull. 8:349. 1927.

Illustrations: Parlatore, 1866, pls. 3, 4; Todaro, 1878, pls. 7, 8, 9; Watt, 1907, pls. 38, 45; Hutchinson, 1947*b*, pl. 8.

Distribution: South America and parts of Central America and the Caribbean, now cosmopolitan in cultivation.

38. *Gossypium mustelinum* Miers ex Watt, Wild and Cult. Cotton Plants 167. 1907.

Gossypium caicoense Aranha et al., Bragantia 28:274. 1969.

Illustrations: Watt, 1907, pl. 26 (reprinted by Mauer, 1954, pl. 125); Valíček, 1974, pl. 77.

Distribution: northeastern Brazil.

39. *Gossypium darwinii* Watt, Wild and Cult. Cotton Plants 68. 1907.

Gossypium barbadense var. *darwinii* (Watt) J. B. Hutchinson, Evol. Gossyp. 51. 1947.

Gossypium barbadense ssp. *darwinii* (Watt) Mauer, Trudy Sredne-Aziatsk. Gosud. Univ. Lenina, n.s., 18(7):24. 1950.

Gossypium barbadense var. *zaria* f. *darwinii* (Watt) Roberty, Candollea 13:91. 1950.

Illustration: Watt, 1907, pl. 4.

Distribution: Galápagos Islands.

The preceding account of *Gossypium* purports only to be an outline, although I hope a useful one. No attempt is made here to deal with the infraspecific classification of the cultivated species, which is a complex topic that is beyond the scope of this work, nor are the lists of synonyms for the cultivated species complete. Moreover, there are many difficulties in determining the identity of individual plants among the tetraploids from the Greater and Lesser Antilles (and elsewhere), where spontaneous and artificial hybridizations among different taxa have blurred the distinctions between them and made a rational classification difficult at best. The key to the tetraploid species (pp. 47–48) may not be (perhaps cannot be) very useful for such interbred materials. For questions on nomenclature or typification, see Fryxell (1976).

The six tetraploid species recognized here constitute a slightly different group than has been included in previous treatments (cf. chapter 6). I accept *G. mustelinum* as a distinct species in part on the basis of incomplete, unpublished data indicating that F_2 genetic breakdown (sterility) is even greater in crosses of *G. mustelinum* with either *G. hirsutum* or *G. barbadense* that it is in F_2 populations of *G. hirsutum* × *G. barbadense*. I accept *G. lanceolatum* as distinct from *G. hirsutum* (and the other tetraploids) in part on the basis of the results reported by Johnson (1975). Further experimental verification of this relationship needs to be made.

Genus *Hampea* Schlechtendal

Hampea Schlechtendal, Linnaea 11:371. 1837; Standley, J. Wash. Acad. Sci. 17:394. 1927; Fryxell, Brittonia 21:359. 1969; Fryxell, Southwestern Nat. 18:465. 1974; Lundell, Wrightia 5:357. 1977.

Trees up to 30 m tall or shrubs, glabrate or invested with stellate hairs, dioecious except for 3 species of section *Standleya*. Leaves petiolate, entire, minutely punctate, simple and narrowly elliptic to broadly cordate or weakly lobed, with 1–9 abaxial foliar nectaries, sometimes basally appendiculate. Stipules usually inconspicuous, linear, usually caducous. Pedicels solitary or more commonly fasciculate in the leaf axils (or on short branched peduncles in *H. micrantha*), 0.5–11.0 cm long, sometimes surmounted by trimerous involucellar nectaries. Involucel of three (or more in *H. tomentosa* and *H. rovirosae*) subulate to linear bracts that are usually inconspicuous and sometimes caducous. Calyx gamophyllous, truncate or rarely obscurely 5-toothed. Flowers odoriferous. Petals yellow-lepidote without where exposed in bud, otherwise white and glabrous, nigro-punctate, reflexed. Staminate flowers with numerous exserted stamens, yellow anthers, no gynoecium; pistillate flowers with a reduced androecium, pallid (non-functional) anthers, an exserted and recurved style, and a decurrent stigma; perfect-flowered species with flowers combining these characteristics. Capsule short-stipitate, 3-loculed (or sometimes 4-loculed), more or less woody, densely green-, yellow-, or brown-puberulent, globose to elongate. Seeds 1–12 per locule, shiny-glabrous, black or red brown, rounded-turbinate, arillate; aril about equaling seed, snow white (when fresh) and soft, quickly dessicating. Embryos punctate, with conduplicate cotyledons. Chromosome number: $2n = 26$.

Type species: *Hampea integerrima* Schlechtendal.

Key to the Species of *Hampea*

A   Leaves simple or lobulate, without basal appendages; corolla without cushion in throat, though petals often pubescent on claw . . . .   B

    Leaves unlobed, basally appendiculate (the appendages sometimes quite small); corolla with dense cushion of white hairs in throat (section *Trianchonia*) . . . .   P

B   Species dioecious; leaves basally cordate, truncate, or cuneate; calyx ecostulate, truncate (or obscurely

5-toothed in *H. tomentosa*); capsules globose or ovoid (nearly as broad as long); seeds 1–4 per locule (except 5–8 in *H. mexicana*); involucellar nectaries usually absent (section *Hampea*) C

Species with perfect flowers; leaves basally cordate; pentamerous nature of calyx often discernible; capsules elongate (about twice as long as broad); seeds 6–12 per locule; involucellar nectaries present (section *Standleya*) .... N

C Leaves glabrate .... D

Leaves puberulent or pubescent .... H

D Capsules brown .... E

Capsules greenish .... F

E Capsule rufous-hirsute within; flowers fasciculate in the axils; foliar nectaries 3–7, distally placed; leaves ovate; stipules 2–10 mm .... 1. *H. integerrima*

Capsule glabrous within; flowers on multiflorate peduncle; foliar nectaries single, basal; leaves lanceolate; stipules 10–12 mm .... 9. *H. micrantha*

F Leaves cordate-ovate, almost as broad as long; pedicels 1–4 cm long; foliar nectaries 5–7, distally placed .....
.... 6. *H. nutricia*

Leaves cuneate-elliptic, more than twice as long as broad; pedicels 3–11 cm long; foliar nectaries 1–3, subbasal .... G

G Fruits 3 cm in diameter, pendulous; seeds 2–3 per locule; leaves about 2.5 times as long as broad, without domatia; corolla about 5 cm in diameter ...... 3. *H. breedlovei*

Fruits 2 cm in diameter, erect; seeds 1 per locule; leaves 2.5–4.0 times as long as broad, with domatia; corolla 2–2.5 cm in diameter .... 2. *H. longipes*

H Foliar nectaries single, basal; leaves often weakly lobed; capsules woody I

Foliar nectaries 3–7; leaves usually simple, capsules coriaceous .... J

I Leaves deeply cordate, with domatia; involucellar nectaries prominent .... 7. *H. tomentosa*

Leaves truncate, lacking domatia; involucellar nectaries lacking .... 8. *H. trilobata*

J Leaves lance-ovate or elliptic; more than twice as long as broad; capsules greenish, white-hirsute within . . . .   K

     Leaves ovate, 1–2 times as long as broad; capsules yellow,
        glabrous within   L

K Involucellar nectaries absent; involucel deciduous; seeds
     solitary . . . . 4. *H. montebellensis*

     Involucellar nectaries present; involucel persistent; seeds
     2 per locule . . . . 5. *H. bracteolata*

L Capsules 1.0–1.5 cm in diameter; seeds 1 per locule;
     petioles weakly puberulent to glabrate . . . .
     . . . . 12. *H. sphaerocarpa*

     Capsules 2.5–3.5 cm in diameter; seeds 2–8 per locule;
     petioles densely puberulent or tomentose . . . .   M

M Leaves twice as long as broad, truncate, simple; foliar
     nectaries 3–7; flowers 1.5–2.0 cm in diameter; seeds
     2–4 per locule; bracts of the involucel 1–3 mm long . .
     . . . . 10. *H. stipitata*

     Leaves 1.0–1.5 times as long as broad, often cordate,
     sometimes slightly lobed; foliar nectaries 1–5; flowers
     2.5–3.0 cm in diameter; seeds 5–8 per locule; bracts
     of the involucel 4–8 mm long . . . . 11. *H. mexicana*

N Leaves puberulent; foliar nectaries 5–9, all distally placed,
     2–8 mm long; calyx transversely rugose . . . .
     . . . . 13. *H. platanifolia*

     Leaves glabrate; foliar nectaries 3–5, at least the lateral
     ones subbasal, 1–2 mm long; calyx not rugose . . . .   O

O Leaf margin ciliate; domatia present; calyx yellowish
     puberulent; flowers 5 cm diameter; capsules apiculate . . .
     . . . . 14. *H. latifolia*

     Leaf margin eciliate; domatia lacking or poorly developed;
     calyx glabrate; flowers 2.5 cm in diameter; capsules
     obtuse or retuse . . . . 15. *H. rovirosae*

P Capsules yellow . . . .   Q

     Capsules brown (or greenish in *H. ovatifolia?*) . . . .   S

Q Leaves often cordate, symmetrically ovate, acuminate,
     1.5–2.0 times as long as broad; flowers 1.5–2.0 cm in
     diameter . . . . 16. *H. thespesioides*

     Leaves more or less truncate, asymmetrical or obtuse;
     flowers 2.5 cm in diameter . . . .   R

R Leaves symmetrical, truncate to slightly cordate, often

obtuse, almost as broad as long, pallid below; involucellar
nectaries lacking; bracts of the involucel often caducous
.... 17. *H. punctulata*
Leaves asymmetrically ovate or elliptic, truncate to
cuneate, 1.5 times as long as broad, brownish below;
involucellar nectaries often present; bracts of the
involucel persistent .... 18. *H. albipetala*
S Foliar nectary single, basal .... 21. *H. ovatifolia*
Foliar nectaries 3–7, distal .... T
T Leaves symmetrically lanceolate or ovate; stipules 3–9 mm
long; bracts of the involucel inserted on the calyx;
calyx hemispheric in bud, usually 5–6 cm long; flowers
1.0–1.5 cm in diameter; filaments 4–5 mm long; capsules 1.0–2.5 cm long...........................
.... 19. *H. appendiculata* var. *appendiculata*
Leaves asymmetrically elliptic or oblong; stipules less than
2 mm long; bracts of the involucel inserted at base of
the calyx; calyx ovoid in bud, usually 10–12 mm long;
flowers 2.5 cm in diameter; filaments 8–11 mm long;
capsules 2–3 cm long ....
.... 20. *H. appendiculata* var. *longicalyx*

Section I. *Hampea*

Series Ia. *Hampea*

1. *Hampea integerrima* Schlechtendal, Linnaea 11:372. 1837.
Distribution: Veracruz, Mexico, between 1,200 m and 1,500 m.

2. *Hampea longipes* Miranda, Ceiba 4:133. 1954. [Fig. 29]
Distribution: Chiapas, Mexico, Guatemala, and Honduras.

3. *Hampea breedlovei* Fryxell, Phytologia 37:291. 1977.
Illustration: Fryxell, Phytologia 37:306. fig. 3. 1977.
Distribution: Chiapas, Mexico.

4. *Hampea montebellensis* Fryxell, Phytologia 37:291. 1977.
Illustration: Fryxell, Phytologia 37:307. fig. 4. 1977.
Distribution: Chiapas, Mexico.

5. *Hampea bracteolata* Lundell, Wrightia 5:357. 1977.
Distribution: Guatemala.

6. *Hampea nutricia* Fryxell, Brittonia 21:372. 1969. [Fig. 30]
Distribution: Veracruz and Tabasco, Mexico, below 1,000 m.

Series Ib. *Preslia* Fryxell

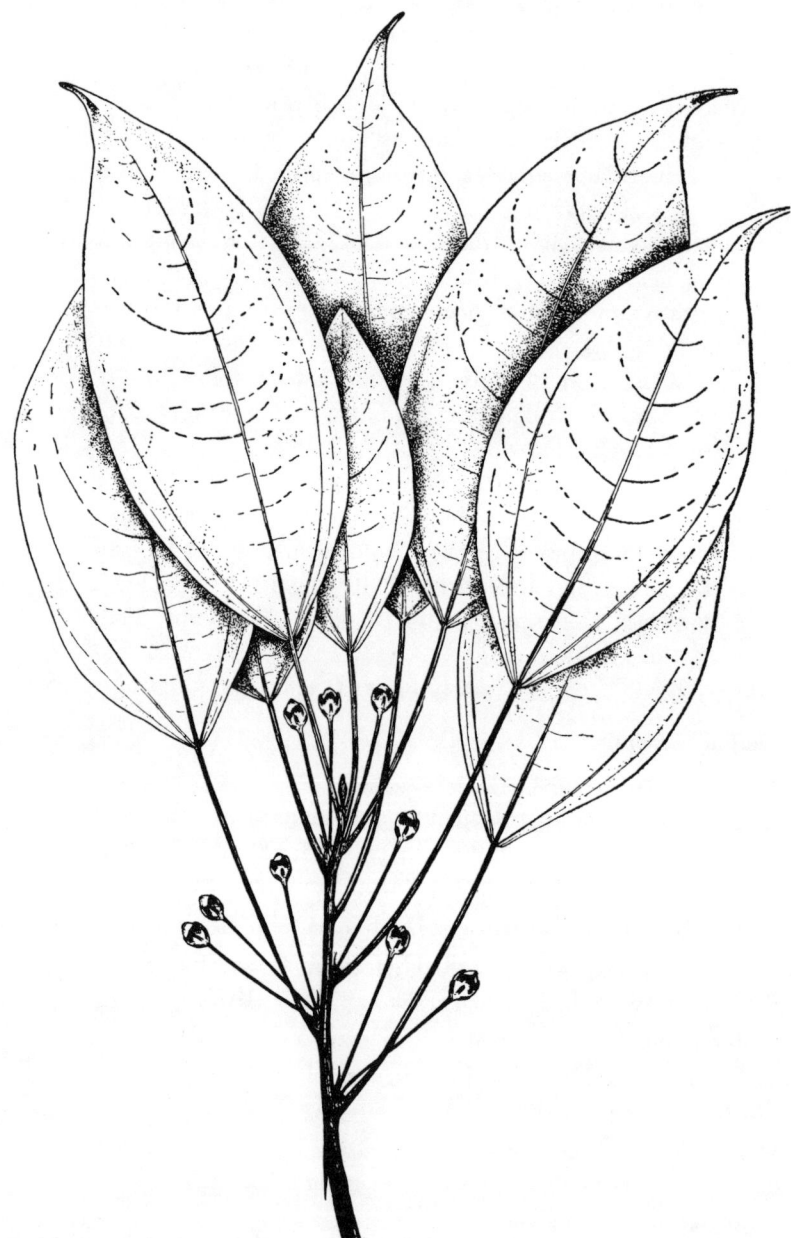

Fig. 29. *Hampea longipes*. (Reprinted from Miranda, Ceiba 4:133. 1954.)

Fig. 30. *Hampea nutricia*. *A*, stem with staminate flowers; *B*, capsules. (Reprinted from T. D. Pennington and J. Sarukhán, *Arboles tropicales de México*. México: Instituto Nacional de Investigaciones Forestales, 1968.)

Preslia Fryxell, Brittonia 21:374. 1969.

Type species: *Hampea tomentosa* (Presl) Standley.

7. *Hampea tomentosa* (Presl) Standley, Contr. U.S. Natl. Herb. 23:787. 1923.

Thespesia tomentosa Presl Rel. Haenk. 2:136. 1836.

Hibiscus preslii Kuntze, Rev. Gen. Pl. 1:69. 1891.

Distribution: Colima, Mexico, and adjacent areas.

8. *Hampea trilobata* Standley, Contr. U.S. Natl. Herb. 23:787. 1923.

Distribution: Yucatán Peninsula (Mexico, British Honduras, and Guatemala).

9. *Hampea micrantha* Robyns, Ann. Missouri Bot. Gard. 55:53. 1968.

Distribution: northeastern Panama.

   Series Ic. *Watsonia* Fryxell

Watsonia Fryxell, Brittonia 21:376. 1969.

Type species: *Hampea stipitata* S. Watson.

10. *Hampea stipitata* S. Watson, Proc. Amer. Acad. Arts 21:460. 1886.

Hampea euryphylla Standley, Publ. Field Mus. Nat. Hist. Bot. Ser. 11:135. 1932.

Distribution: principally Guatemala, but also adjacent areas of British Honduras, El Salvador, and Mexico (Chiapas).

11. *Hampea mexicana* Fryxell, Brittonia 21:380. 1969.

Distribution: Oaxaca and Chiapas, Mexico, and Guatemala.

12. *Hampea sphaerocarpa* Fryxell, Brittonia 21:380. 1969.

Distribution: northern Honduras and extreme eastern Guatemala.

Section II. *Standleya* Fryxell

Standleya Fryxell, Brittonia 21:382. 1969.

Type species: *Hampea platanifolia* Standley.

13. *Hampea platanifolia* Standley, J. Wash. Acad. Sci. 17:317. 1927.

Distribution: Costa Rica and Nicaragua.

14. *Hampea latifolia* Standley, Pub. Field Mus. Nat. Hist. Bot. Ser. 22:90. 1940.

Distribution: southwestern Guatemala and adjacent Chiapas on the Pacific coast.

15. *Hampea rovirosae* Standley, J. Wash. Acad. Sci. 17:397. 1927.

Hampea macrocarpa Lundell, Lloydia 2:102. 1939.

Distribution: northwestern Guatemala; Tabasco, Mexico; and disjunct in southern Chiapas, Mexico.

Section III. *Trianchonia* Fryxell

Trianchonia Fryxell, Brittonia 21:386. 1969.

Type species: *Hampea thespesioides* Triana & Planchon.

16. *Hampea thespesioides* Triana & Planchon, Ann. Sci. Nat. Bot., ser. 4, 17:188. 1862.

Distribution: central Colombia, above 1,000 m.

17. *Hampea punctulata* Cuatrecasas, Phytologia 4:472. 1954.

Hampea romeroi Cuatrecasas, *loc. cit.*

Hampea dukei Robyns, Ann. Missouri Bot. Gard. 55:52. 1968.

Distribution: lowlands of the Panama-Colombia border.

18. *Hampea albipetala* Cuatrecasas, Lloydia 11:192. 1949.

Distribution: Colombia, in the Dept. del Valle, to Panama.

19. *Hampea appendiculata* (J. Donnell-Smith) Standley, J. Wash. Acad. Sci. 17:395. 1927.

Hampea integerrima var. *appendiculata* J. Donnell-Smith, Bot. Gaz. 27:331. 1899.

Hampea panamensis Standley, *op. cit.*, 396.

Distribution: Costa Rica and western Panama.

20. *Hampea appendiculata* (J. Donnell-Smith) Standley var. *longicalyx* Fryxell, Brittonia 21:391. 1969.

Distribution: eastern Panama.

21. *Hampea ovatifolia* Lundell, Wrightia 4:140. 1970.

Distribution: Guatemala.

Descriptions, illustrations, distributional data, specimen citations, and other details may be found in the monograph of the genus (Fryxell, 1969; see also Fryxell, 1974).

Genus *Kokia* Lewton

Kokia Lewton, Smithson. Misc. Coll. 60(5):2. 1912; Rock, Bot. Bull. Haw. Board Agr. For. 6:1. 1919; Degener, Fl. Haw., Fam. 221. 1934; Hutchinson, New Phyto!. 46:138. 1947.

Trees up to 10 m tall. Leaves usually cordate, shallowly palmately 7-lobed, suborbicular in total outline, abundantly punctate, glabrous or sparsely pubescent, sometimes densely woolly at juncture of petiole and leaf blade; lobes subequal, triangular, obtuse or rounded-acute. Foliar nectaries lacking. Petioles as long as or longer than lamina, terete, punctate, glabrous throughout or densely pubescent at distal and proximal ends. Stipules falcate, 3–5 mm long, 1–2 mm wide, glabrous or woolly, punctate, caducous. Peduncles axillary, stout, jointed, bracteate at articulation, surmounted by an involucel of 3 bracts. Involucellar nectaries lacking. Bracts distinct, lanceolate to broadly and irregularly ovate or orbicular, sessile, entire, sinuate or irregularly lobed, accrescent, recticulate-veined, 2–3 times as long as calyx. Calyx tube urceolate or cylindrical, 5-lobed, "often with a median transverse vein, the upper half of the calyx usually breaking off at this point, giving the calyx the appearance of being truncate" (Lewton), glabrous or pubescent. Petals red, glabrous within, yellowish-pubescent without , exceeding bracteoles, curved in bud, reflexed in flower. Staminal column dark red, curved, antheriferous throughout most of length; filaments recurved. Style and stigma clavate. Capsules 5-celled, woody, ovoid, dehiscing tardily. Seeds one per carpel, densely lanate, up to 2 cm long; seed hairs reddish. Embryos punctate with conduplicate cotyledons. Chromosome number: $2n = 24$.

Type species: *Kokia rockii* Lewton (= *K. drynarioides*).

Key to the Species of *Kokia*

A   Bracts of the involucel at least twice as long as broad; leaves less than 8 cm long, subtruncate with broadly open basal sinus .... 1. *K. lanceolata*

Bracts of the involucel approximately as broad as long; leaves usually longer than 8 cm, deeply cordate with a very narrow basal sinus or the sinus closed by the overlapping of the basal lobes .... B

B   Foliage essentially glabrous, hairs essentially confined to the axils of the main veins of the lower leaf surface; seeds 1 cm long .... 2. *K. cookei*

Proximal and distal ends of petioles, and base of lamina (below) near attachment of petiole, densely woolly; seeds 2 cm long C

C   Calyx more or less glabrous without, but with a ring of stiff brownish hairs within at the point of insertion of the petals; flowers up to 15 cm in diameter; capsules up to 3.0 cm long: .... 3. *K. drynarioides*

Calyx densely yellowish-hirsute without, but lacking a ring of hairs within; flowers up to 22 cm in diameter; capsules up to 3.7 cm long: .... 4. *K. kauaiensis*

1. *Kokia lanceolata* Lewton, Smiths. Misc. Coll 60(5):4. 1912.
Gossypium drynarioides var. β. Hillebrand, Fl. Haw. Isl. 51. 1888.
Kokia drynarioides var. *lanceolata* (Lewton) Rock, Indig. Trees Haw. Isld. 307. 1913.
Distribution: eastern Oahu, Hawaii; now extinct.

Fig. 31. *Kokia drynarioides*. (Reprinted from Degener, 1965.)

2. *Kokia cookei* Degener, Fl. Haw., Fam. 221. 1934.

Gossypium drynarioides sensu Hillebrand, Fl. Haw. Isl. 51. 1888 (non Seemann, Fl. Vit. 22. 1865).

Distribution: western Molokai, Hawaii; extinct in nature but surviving in cultivation.

3. *Kokia drynarioides* (Seemann) Lewton, Smiths. Misc. Coll. 60 (5):3, 1912. [Fig. 31]

Gossypium drynarioides Seemann, Fl. Vit. 22. 1865.
Hibiscus drynarioides (Seemann) Kuntze, Rev. Gen. Pl. 1:68, 1891.
Kokia rockii Lewton, *loc. cit.*
Distribution: North Kona District, Island of Hawaii; probably extinct in nature, but surviving in cultivation.

4. *Kokia kauaiensis* (Rock) Degener & Duvel in Degener, Fl. Haw., Fam. 221. 1934. [Fig. 32]

Kokia rockii var. *kauaiensis* Rock, Bot. Bull. Haw. Board Agr. For. 6:16, 1919.
Distribution: western Kauai, Hawaii.

Fig. 32. *Kokia kauaiensis*. (Reprinted from Degener, 1934.)

Genus *Lebronnecia* Fosberg

Lebronnecia Fosberg in Fosberg & Sachet, Adansonia, n.s., 6:509. 1966.

Small tree to 9 m tall. Bark brownish. Leaves simple, cordate, broadly ovate, acuminate, slightly longer than broad, entire, palmately 7-nerved (the nerves prominent below and densely and minutely nigro-punctate), sparsely stellate-pubescent in meristematic tissue, becoming quickly and completely glabrate, abundantly punctate throughout. Foliar nectary single, subbasal, elongate, often obscure in dried specimens (because it is simply a secretory zone on the surface of the midrib, not a morphologically defined structure). Petioles equaling or somewhat shorter than lamina, glabrate, terete, punctate. Stipules 3–5 mm long, filiform, punctate, caducous. Inflorescence an axillary sympodium, usually reduced to 1 or 2 flowers (any additional distal buds usually shriveling and abscising at a very early stage); each pedicel glabrous, articulated on the peduncle and surmounted by an involucel; involucellar nectaries lacking; bracts of the involucel 3, irregularly inserted, shorter than the calyx, linear-acute, glabrous. Calyx gamophyllous, 7–15 mm long, 5-toothed (teeth 1–2 mm long), 5-ribbed, punctate, glabrate, splitting to the base in flower to pseudobilabiate form. Petals white with tinge of pink in throat, fading brownish, 3 cm long, densely punctate, yellowish-pubescent without (where exposed in bud), otherwise glabrous, lacking basal spot, narrowly and asymmetrically obovate, distal portion of blade crinkled, with acute tooth on outer margin distally. Staminal column 12–13 mm long, pallid, punctate; anthers numerous (about 100), subsessile, pallid. Stigma barely exceeding androecium, the lobes decurrent. Ovary 3-celled, greenish-puberulent, ovoid. Capsule subspherical to obovoid, yellowish puberulent, 2–3 cm long, glabrous within, the suture not or scarcely ciliate; carpel wall 1–2 mm thick. Seeds 1 per locule, almost 1.5 cm long, densely covered with straight red brown hairs 1 cm long. Embryo punctate with conduplicate cotyledons enclosing mesocotyl and hypocotyl. Chromosome number not known.

Type species: *Lebronnecia kokioides* Fosberg.

Genus monotypic.

1. *Lebronnecia kokioides* Fosberg, Adansonia, n.s., 6:510. 1966.
Characters of the genus.
Illustrations: Fryxell, 1968, fig. 6*a*; D'Arcy, 1976 [unpaged].
Distribution: Marquesas Islands.

Genus *Thespesia* Solander

Thespesia Solander ex Correa, Ann. Mus. Paris 9:290. 1807, nom. conserv.;
  Baker, J. Bot. 35:50. 1897; Hutchinson, New Phytol. 46:134. 1947; How-
  ard, Bull. Torrey Bot. Club 76:89. 1949; Exell & Hillcoat, Estud., Ensai.
  & Docum. 12:58. 1954; Fryxell, Bot. Gaz. 129:296. 1968; Fosberg &
  Sachet, Smiths. Contr. Bot. No. 7:7. 1972.
Bupariti Duhamel, Sem. Pl. Arbr., Add. 5. 1760, nom. rejic.
Montezuma Mociño & Sessé ex DC., Prodr. 1:477. 1824.
Azanza Alefeld, Bot. Zeit. 19:298. 1861 (non Mociño & Sessé inedit.)
Maga Urban, Symb. Antill. 7:281. 1912.
Ulbrichia Urban, Dansk Bot. Arkiv 4(7):7. 1924.
Shantzia Lewton, J. Wash. Acad. Sci. 18:15. 1928.
Armouria Lewton, *op. cit.* 23:64. 1933.
Atkinsia Howard, Bull. Torrey Bot. Club 76:97. 1949.
Thespesiopsis Exell & Hillcoat, Estud., Ensai. & Docum. 12:55. 1954.

Shrubs or more commonly trees, often evergreen, the foliage usually punctate, glabrous or with peltate or stellate indumentum. Leaves
petiolate, simple or shallowly lobed, cordate to subtruncate, ovate or
trilobulate, entire, obtuse to acuminate, palmately or pedately 5- to
9-nerved (rarely penninerved). Stipules subulate or filiform (absent in
T. cubensis), caducous. Flowers large and showy, borne singly on
axillary peduncles (sometimes aggregated terminally); pedicels sometimes subtended by 2-merous bractlets and surmounted by an involucel of 3 to many elements that are either whorled or irregularly
(spirally) inserted and often caducous; 3-merous involucellar nectaries
sometimes present. Calyx truncate to 5-lobed, persistent (or cir-
cumscissilly deciduous in *T. grandiflora*). Petals white, yellow, or
rose, with or without a dark spot on claw, often crinkled distally.
Staminal column long, pallid; anthers numerous. Style and stigma
stout, clavate; stigmas decurrent, sometimes twisted. Capsules (3-)5-
celled, dehiscent or indehiscent, coriaceous or ligneous. Seeds turbinate (sometimes angularly so), glabrous or pubescent, large (7–13 mm
long); embryos punctate, with conduplicate cotyledons enclosing the
mesocotyl and hypocotyl.

Chromosome number: $2n = 26$; counts of 24 and 28 are also reported.

Type species: *Thespesia populnea* (L.) Solander ex Correa.

Key to the Species of *Thespesia*

A Calyx truncate or at most 5-denticulate; bracts of the in-
      volucel 3 to many, if numerous then evenly distributed
      around base of calyx; foliage pubescent, lepidote, or
      glabrate; capsules dehiscent or indehiscent, pubescent,
      lepidote, or glabrate (section *Thespesia*) . . . .   B

Calyx 5-lobed; 2(–3) subulate bractlets above each of 3 involucral nectaries producing an involucel of 6(–9) minute bractlets that are clearly grouped; foliage discolorous, densely pubescent or puberulent; capsules dehiscent, pubescent or puberulent (section *Lampas*) ..... P

B Bracts of the involucel 3, whorled or spirally disposed; involucellar nectaries often present C

Bracts of the involucel 6 or more, whorled, sometimes caducous before anthesis; involucellar nectaries absent.... K

C Leaves strongly discolorous, broader than long, shallowly 3-lobed, deeply cordate with open, obtuse basal sinus and pedate nervation; petals 8 cm long, white fading brownish .... 8. *T. beatensis*

Leaves concolorous, usually longer than broad, simple (or 3-lobed in *T. acutiloba*, but basally truncate), truncate or cordate with more or less acute basal sinus and palmate nervation; petals 3–11 cm long, yellow or red .... D

D Petals red, 7–11 cm long; calyx circumscissile and deciduous in fruit .... 9. *T. grandiflora*

Petals yellow often with dark red spot at base, 3–7 cm long; calyx persistent in fruit.............................. E

E Bracts of the involucel to 6 mm broad, inserted on calyx, persistent in fruit .... 7. *T. danis*

Bracts of the involucel narrower, inserted at base of calyx or on pedicel, usually caducous .... F

F Flowers and fruits congested in terminal aggregations ..... G

Flowers and fruits solitary in the axils, scattered .... H

G Staminal column exceeding corolla; domatia well developed; stipules absent; fruits globose, apiculate .... 5. *T. cubensis*

Staminal column included in corolla; domatia poorly developed or absent; stipules minute, caducous; fruits depressed, 5-lobed .... 6. *T. mossambicensis*

H Leaves basally truncate, trilobed .... 4. *T. acutiloba*

Leaves more or less cordate, simple I

I Indumentum of minute stellate hairs.... 3. *T. patellifera*

Indumentum of minute lepidote scales or glabrate .... J

J Leaves green, deeply cordate; pedicels erect, 1–5 cm long,

with a bibracteate articulation near base . . . .
. . . . .1. *T. populnea*
Leaves copper-colored, shallowly cordate; pedicels often
   drooping, 5–12 cm long, without bracteate articulation
   . . . . 2. *T. populneoides*
K Peduncles articulated well above middle, bibracteate at
   articulation; involucel usually persistent in fruit   L
   Peduncles articulated at middle or below (sometimes at
   very base and appearing inarticulate); involucel usually
   caducous . . . .   N
L Petals 5–6 cm long; bracts of the involucel 18–22, persis-
   tent in fruit; leaves cordate, palmately nerved . . . .
   . . . . 10. *T. multibracteata*
   Petals 3–4 cm long; bracts of the involucel 6–10; leaves
   truncate or weakly cordate . . . .   M
M Bracts of the involucel 6–8, persistent in fruit; leaves
   appearing penninerved . . . . 11. *T. fissicalyx*
   Bracts of the involucel 8–10, caducous before anthesis;
   leaves palmately nerved . . . . 12. *T. robusta*
N Leaves densely pubescent, trilobed . . . . . 13. *T. garckeana*
   Leaves lepidote to glabrate, usually simple . . . .   O
O Leaves truncate, simple; stigma lobes and carpels 3–4
   . . . . 14. *T. gummiflua*
   Leaves cordate, sometimes trilobed; stigma lobes and car-
   pels 5 . . . . : . . . . 15. *T. trilobata*
P Leaves usually simple, acute to obtuse, densely puberulent
   below, the stellate hairs to 0.25 mm in diameter; seeds
   pubescent . . . . 17. *T. thespesioides*
   Leaves usually trilobed, acute to acuminate, densely
   pubescent below, the hairs to 1.0 mm in diameter;
   seeds glabrous . . . . 16. *T. lampas*

Section I. *Thespesia*

1. *Thespesia populnea* (L.) Solander ex Correa, Ann. Mus. Par. 9:290.
   1807. [Fig. 33]
Hibiscus populneus L., Spec. Pl. 694. 1753.
Hibiscus bacciferus G. Forster, Prodr. 48. 1786.
Malvaviscus populneus (L.) Gaertner, Fruct. et Sem. 2:253, pl. 135, f. 3. 1791.
Hibiscus populifolius Salisbury, Prodr. 383. 1796.
Thespesia macrophylla Blume, Bijdr. 2:73. 1825.

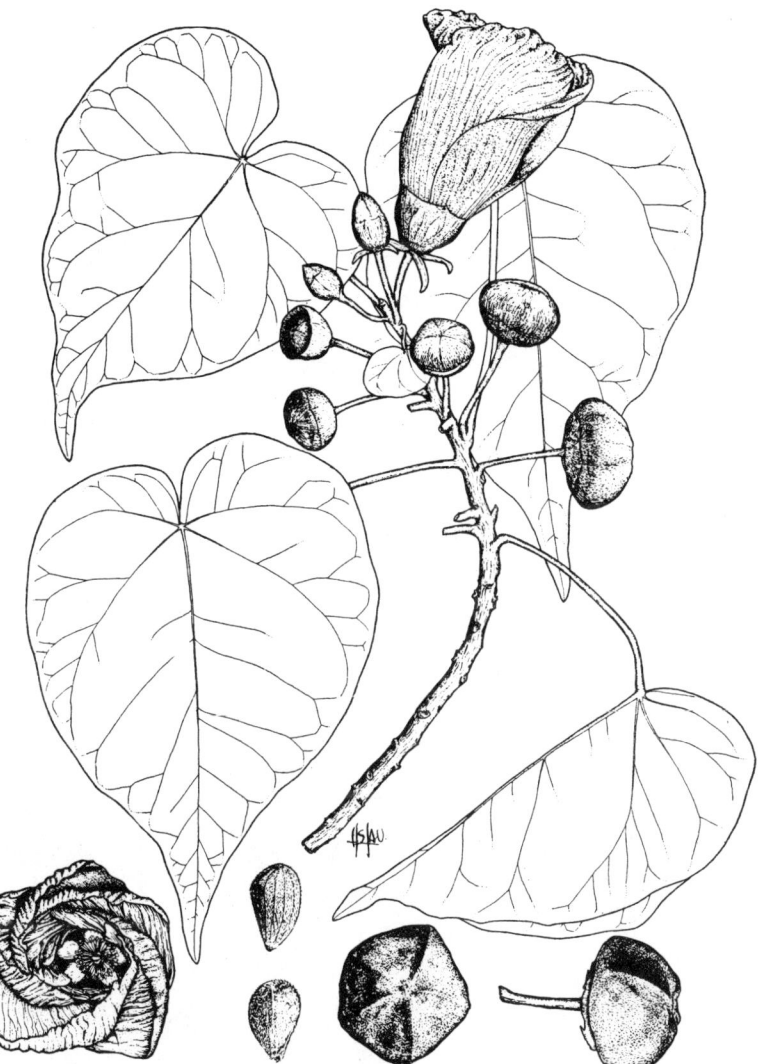

Fig. 33. *Thespesia populnea.* (Reprinted from Degener, 1932.)

Hibiscus blumei Kuntze, Rev. Gen. Pl. 1:69. 1891.

Illustrations: Pierre, 1888, pl. 173A; Li, 1963, fig. 214.

Distribution: pantropical in littoral habitats; often planted as an ornamental shade tree in the tropics.

2. *Thespesia populneoides* (Roxburgh) Kosteletzky, Allg. Med. Pharm. Fl. 5:1861. 1836. [Fig. 34]

Thespesia populnea sensu Blume, Bijdr. 2:73. 1825 (non [L.] Solander ex Correa).

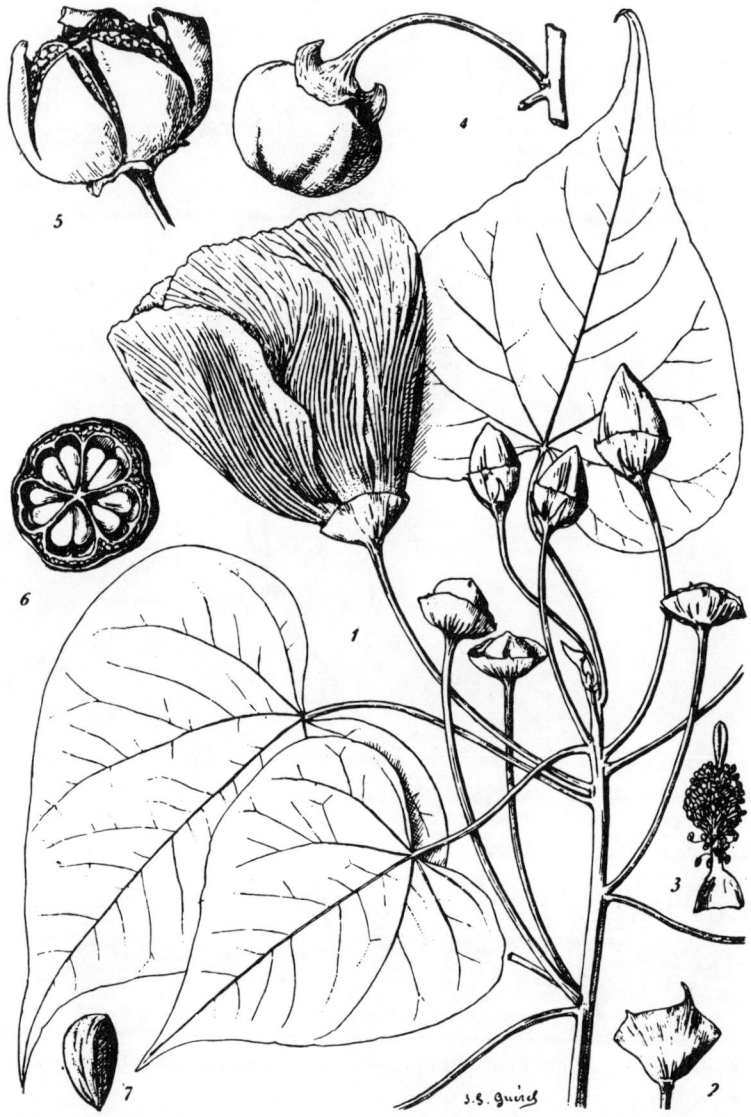

Fig. 34. *Thespesia populneoides.* 1, flowering branch (× 2/3); 2, form of the calyx with bract scars (× 2/3); 3, staminal column and style (× 2/3); 4, fruit (× 2/3); 5, fruit at dehiscence (× 2/3); 6, cross-section of fruit (× 2/3); 7, seed (× 3/2). (Reprinted from Hochreutiner, 1955, pl. 30; given there as *Thespesia populnea.*)

Hibiscus populneoides Roxburgh, Fl. Ind. 3:181. 1832.

Thespesia banalo Blanco, Fl. Filip., ed. II., 382. 1845.

Thespesia populnea var. *populneoides* (Roxburgh) Pierre, Fl. Forest. Cochinch. 3: pl. 173B. 1888.

Thespesia howii Hu, Fl. China, Fam. 153:69, t. 22, f. 3. 1955.

Illustrations: Pierre, 188, pl. 173B; Hochreutiner, 1955, pl. 30; Hu, 1955.

Distribution: shores of the Indian Ocean and its islands; sparingly cultivated elsewhere.

3. *Thespesia patellifera* Borssum Waalkes, Blumea 4 (suppl.):154. 1958.

Illustration: van Borssum Waalkes, Blumea 4 (suppl.):154, fig. 2. 1958.

Distribution: New Guinea.

4. *Thespesia acutiloba* (E. G. Baker) Exell & Mendonça, Estud., Ensai. & Docum. 12:63. 1954. [Fig. 35]

Thespesia populnea var. *acutiloba* E. G. Baker, J. Bot. 35:51. 1897.

Illustrations: Exell, 1961, pl. 82; Verdoorn, Fl. Plants Africa 37: pl. 1468. 1966.

Distribution: southern Mozambique, Natal.

5. *Thespesia cubensis* (Britton & Wilson) J. B. Hutchinson, New Phytol. 46:135. 1947. [Fig. 36]

Maga cubensis Britton & Wilson, Mem. Torrey Bot. Club 16:81. 1920.

Montezuma cubensis (Britton & Wilson) Urban, Repert. Sp. Nov. 18:117. 1922.

Atkinsia cubensis (Britton & Wilson) Howard, Bull. Torrey Bot. Club 76:97. 1949.

Illustration: Howard, 1949, figs. 4–14.

Distribution: Cuba.

6. *Thespesia mossambicensis* (Exell & Hillcoat) Fryxell, Bot. Gaz. 129:301. 1968. [Fig. 37]

Thespesiopsis mossambicensis Exell & Hillcoat, Estud., Ensai. & Docum. 12:55. 1954.

Illustrations: Exell and Hillcoat, 1954, pl. 8; Exell, 1961, pl. 83.

Distribution: northern Mozambique.

7. *Thespesia danis* Oliver in Hooker, Icon. Pl., t. 1336. 1881.

Illustration: Oliver, in Hooker, Icon. Pl., t. 1336. 1881.

Distribution: East Africa.

8. *Thespesia beatensis* (Urban) Fryxell, Bot. Gaz. 129:301. 1968.

Ulbrichia beatensis Urban, Dansk Bot. Arkiv 4:8. 1924.

Armouria beata Lewton, J. Wash. Acad. Sci. 23:64. 1933.

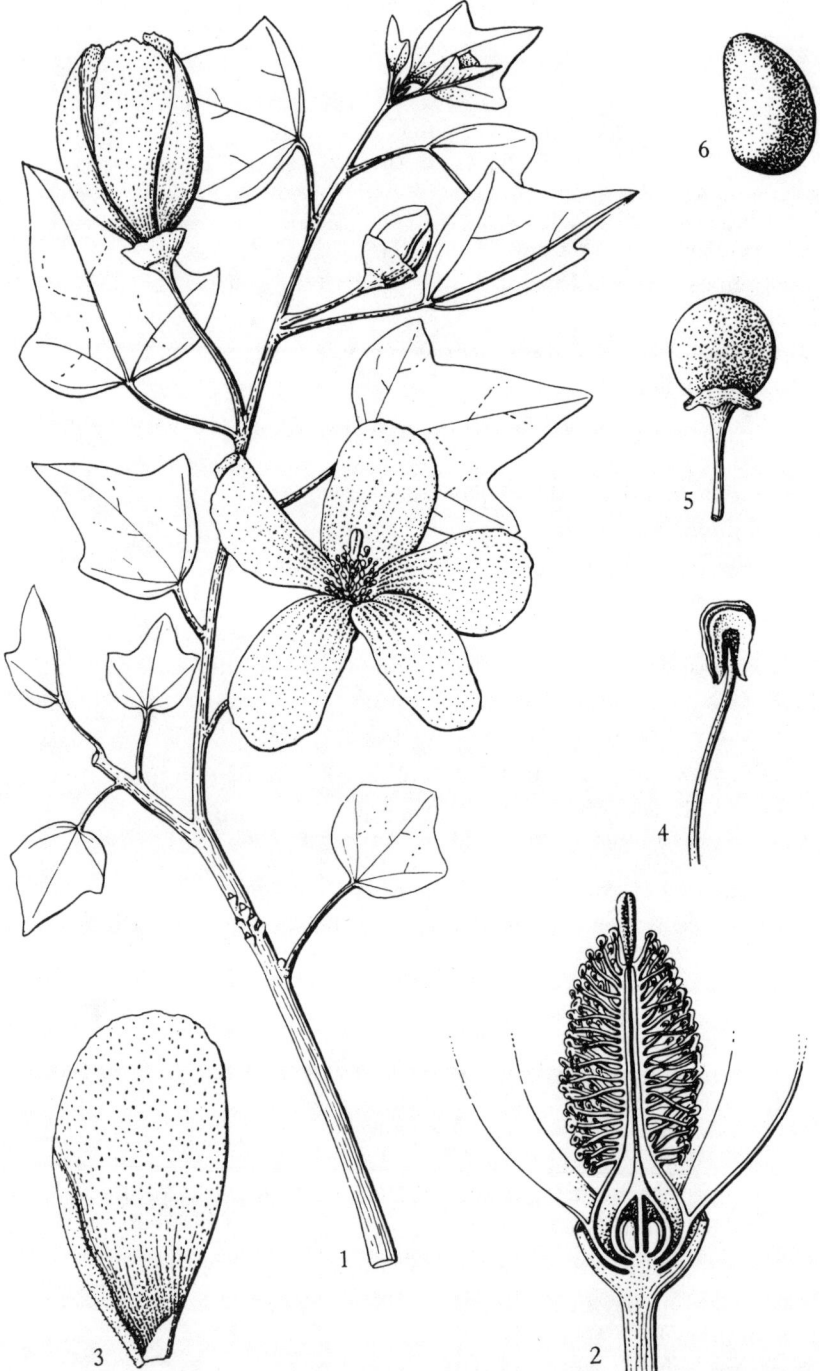

Fig. 35. *Thespesia acutiloba*. 1, flowering branch (× 2/3); 2, vertical section of androecium and gynoecium (× 2); 3, petal (× 1); 4, stamen (× 6); 5, fruit (× 2/3); 6, seed (× 2). (Reprinted from Exell, 1961, pl. 82, by courtesy of the editorial board, *Flora Zambesiaca*.)

Fig. 36. *Thespesia cubensis.* 7, habit, showing leafless cluster of flowers and persistent woody calyx (pubescence omitted); 8, bud; 9, cross-section of fruit with two fertile locules; 10, cross-section of seed, showing black, punctate cotyledons; 11, longitudinal section of seed; 12, side view of seed; 13, lower surface of leaf blade, showing axillary webs and scale insect domatia; 14, flower, showing staminal column exceeding the corolla. All drawings × 1/2. (Reprinted from Howard, 1949, figs. 7–14; given there as *Atkinsia cubensis.*)

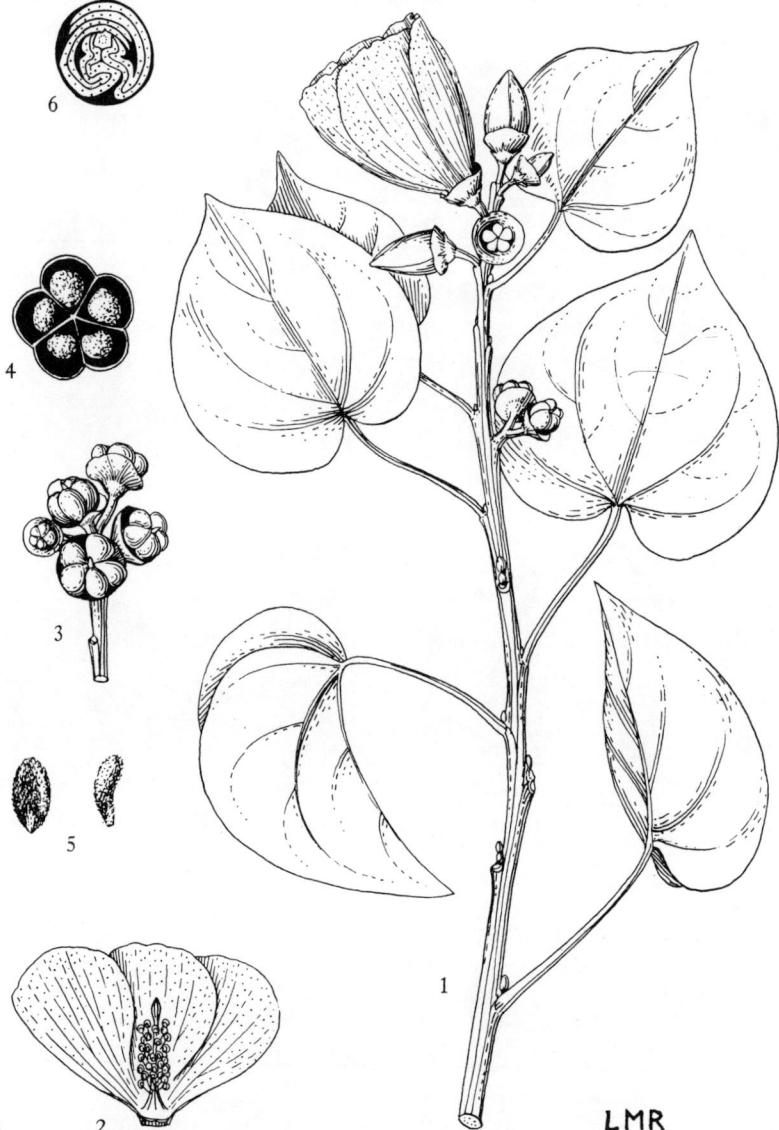

LMR

Fig. 37. *Thespesia mossambicensis.* 1, flowering and fruiting branch (× 2/3); 2, flower with calyx and two petals removed (× 2/3); 3, infructescence (× 2/3); 4, cross-section of fruit (× 4/3); 5, seed (× 3/4); 6, cross-section of seed (× 2). (Reprinted from Exell, 1961, pl. 83, by courtesy of the editorial board, *Flora Zambesiaca;* given there as *Thespesiopsis mossambicensis.* After a drawing by Miss D. Hillcoat.)

Thespesia beata (Lewton) J. B. Hutchinson, New Phytol. 46:136. 1947.

Illustration: Urban, 1924, pl. 1.

Distribution: endemic to Isla Beata, Dominican Republic.

9. *Thespesia grandiflora* DC., Prodr. 1:456. 1824. [Fig. 38]

Montezuma speciosissima Sessé & Mociño ex DC., *op. cit.*, 477.

Maga grandiflora (DC.) Urban, Symb. Antil. 7:281. 1912.

Montezuma grandiflora (DC.) Urban ex Urban & Helwig, Repert. Spec. Nov. 24:238. 1928.

Illustrations: Howard, 1949, figs. 1–6; Menninger, 1962, pl. 195; Little and Wadsworth, 1964, pl. 151.

Distribution: native to Puerto Rico, but planted as ornamental tree in other parts of the neotropics.

10. *Thespesia multibracteata* Borssum Waalkes, Blumea 14:114. 1966. [Fig. 39]

Illustration: van Borssum Waalkes, 1966, pl. 16.

Distribution: New Guinea.

11. *Thespesia fissicalyx* Borssum Waalkes, Blumea 14:112. 1966. [Fig. 40]

Illustration: van Borssum Waalkes, 1966, pl. 15.

Distribution: New Guinea.

12. *Thespesia robusta* Borssum Waalkes, Blumea 14:111. 1966. [Fig. 41]

Illustrations: van Borssum Waalkes, 1966, pl. 14*a, b*.

Distribution: New Guinea.

13. *Thespesia garckeana* F. Hoffmann, Beitr. Kennt. Fl. Centr.-Ost-Afr. 12. 1889. [Fig. 42]

Thespesia debeertsii DeWildemann & Durand, Ann. Mus. Congo Belge Ser. Bot. 2, 1(2):6. 1898.

Thespesia hockii DeWildemann, Bull. Jard. Bot. Brux. 3:266. 1911.

Thespesia rogersii S. Moore, J. Bot. 56:5. 1918.

Shantzia garckeana (F. Hoffmann) Lewton, J. Wash. Acad. Sci. 18:13. 1928.

Azanza garckeana (F. Hoffmann) Exell & Hillcoat, Estud., Ensai. & Docum. 12:59. 1954.

Illustrations: Lewton, 1928, pls. 1, 2; Exell, 1961, pl. 88.

Distribution: southeastern Africa.

14. *Thespesia gummiflua* Capuron, Adansonia, n.s., 8:7. 1968. [Fig. 43].

Illustration: Capuron, 1968, pl. 1.

Distribution: Madagascar.

15. *Thespesia trilobata* E. G. Baker, J. Bot. 35:52. 1897.

Distribution: East Africa.

Fig. 38. *Thespesia grandiflora*. 1, habit with pubescence omitted (note fruiting pedicel after capsule has fallen); 2, glabrous seed; 3, circumscissile calyx and young fruiting pedicel; 4, longitudinal section of fruit; 5, mature fruit; 6, cross-section of mature fruit showing plump seeds in four locules. All drawings × 1/2. (Reprinted from Howard, 1949, figs. 1–6; given there as *Montezuma grandiflora*.)

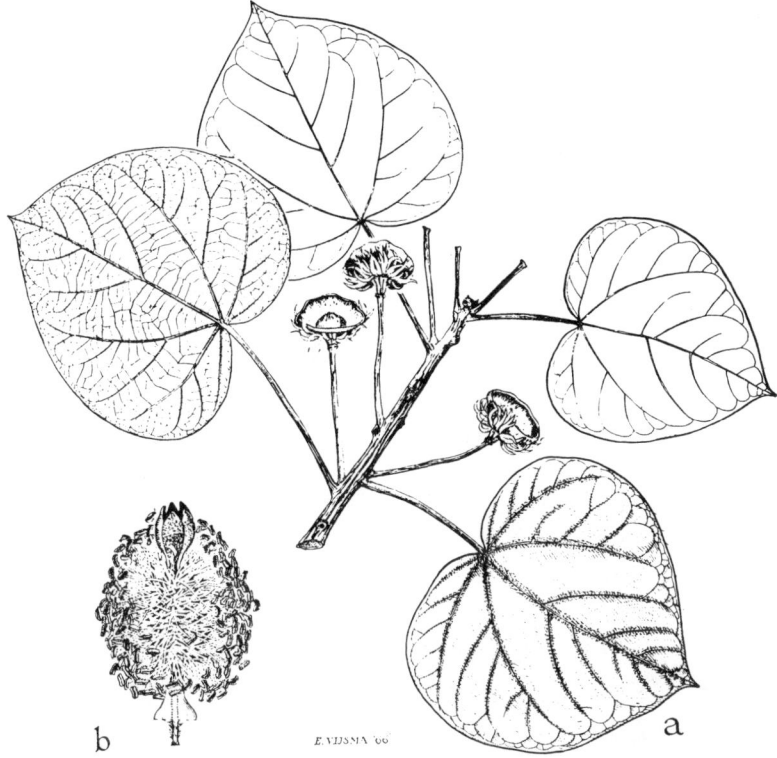

Fig. 39. *Thespesia multibracteata*. *a*, habit (× 1/3); *b*, staminal column and style (× 1). (Reprinted from Borssum Waalkes, 1966, pl. 16.)

Section II. *Lampas* (Ulbrich) Borssum Waalkes

Lampas (Ulbrich) Borssum Waalkes, Blumea 14:115. 1966.

Basionym: *Hibiscus* sect. *Lampas* Ulbrich, Notzbl. Bot. Gart. Berlin-Dahlem 8:158. 1922.

Type species: *Thespesia lampas* (Cavanilles) Dalzell ex Dalzell & Gibson.

16. *Thespesia lampas* (Cavanilles) Dalzell ex Dalzell & Gibson, Bombay Fl. 19. 1861.

Hibiscus lampas Cavanilles, Diss. 3:154. 1787.

Hibiscus callosus Blume, Bijdr. 2:67. 1825.

Paritium gangeticum (Roxburgh) G. Don, Gen. Hist. 1:485. 1831.

Hibiscus tetralocularis Roxburgh, Hort. Beng. 97. 1814; Fl. Ind. 3:198. 1832.

Hibiscus gangeticus Roxburgh ex Wight & Arnott, Prodr. 49. 1834.

Thespesia sublobata Blanco, Fl. Filip., ed. II, 382. 1845.

Azanza lampas (Cavanilles) Alefeld, Bot. Zeit. 19:298. 1861.

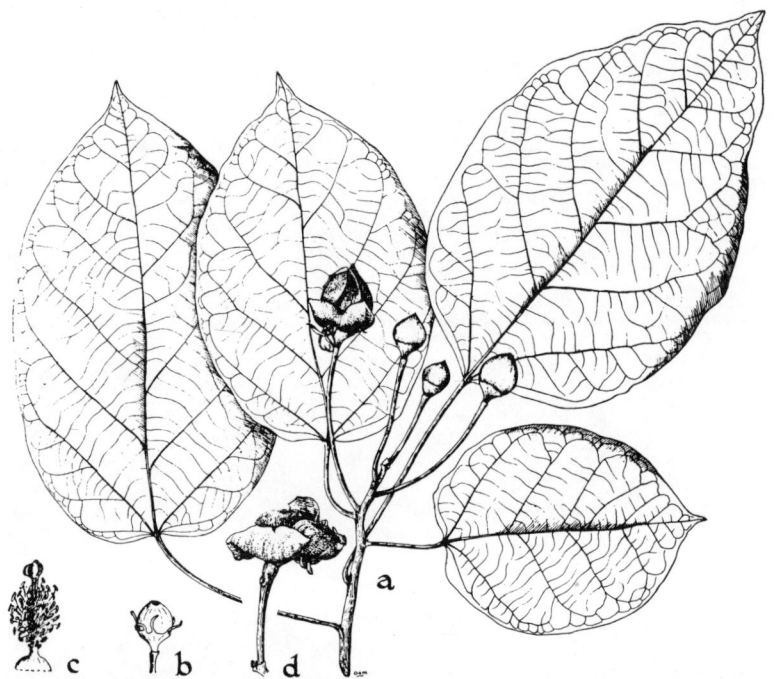

Fig. 40. *Thespesia fissicalyx. a*, habit (× 1/2); *b*, bud (× 1/2); *c*, staminal column (× 1); *d*, fruit (× 1/2). (Reprinted from Borssum Waalkes, 1966, pl. 15.)

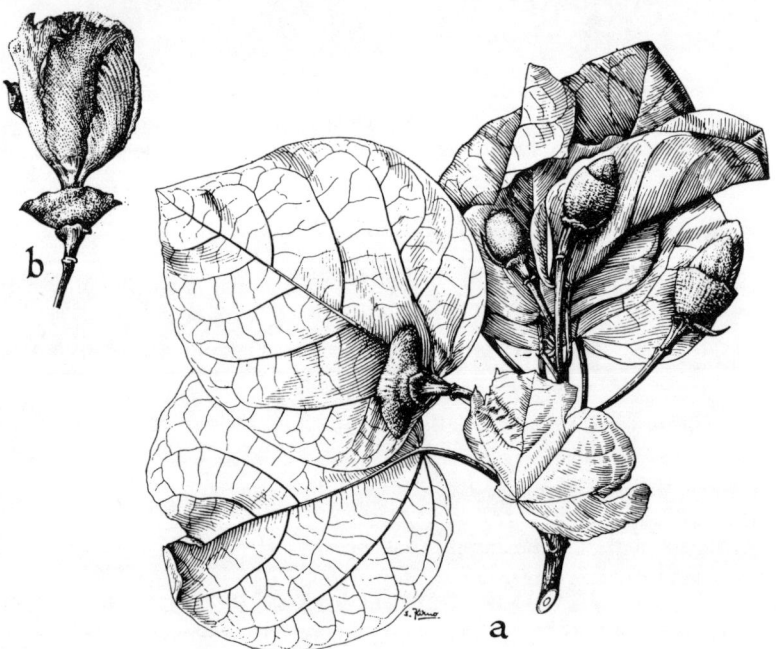

Fig. 41. *Thespesia robusta. a*, habit (× 1/2); *b*, flower (× 1/2). (Reprinted from Borssum Waalkes, 1966, pl. 14.)

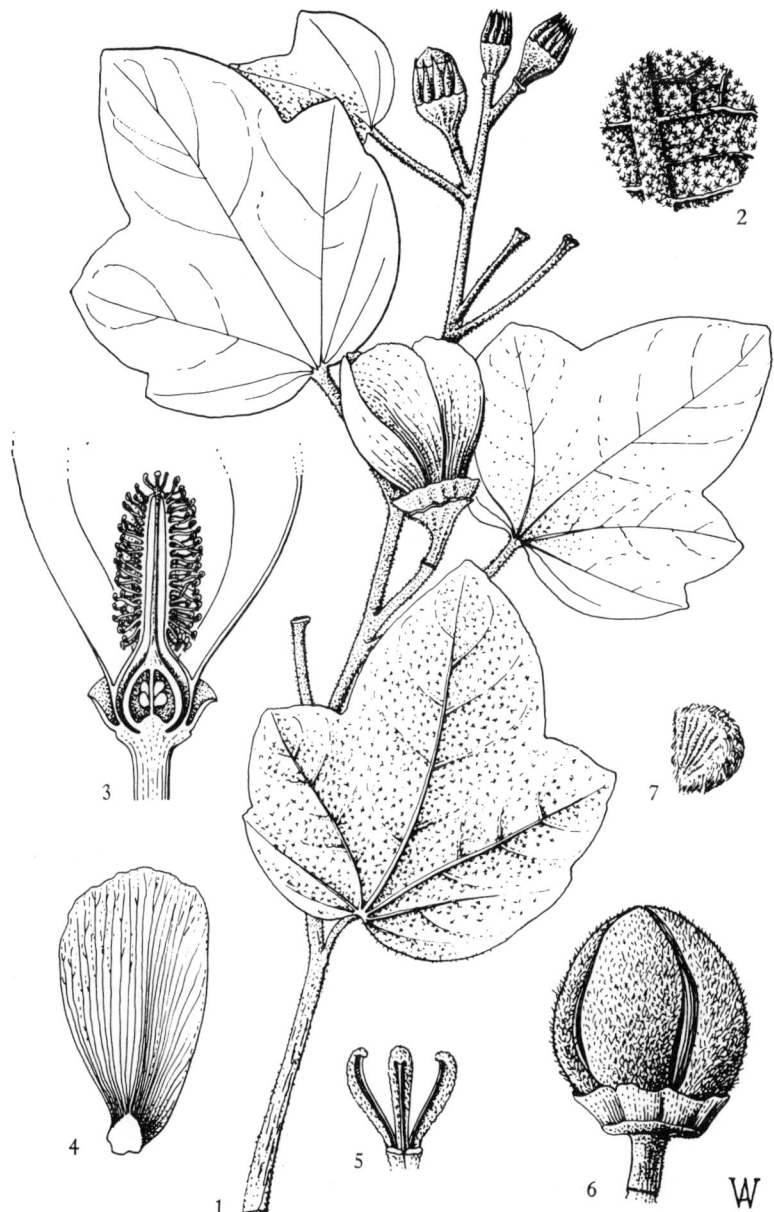

Fig. 42. *Thespesia garckeana.* 1, flowering branch (× 2/3); 2, part of under surface of leaf (× 4); 3, vertical section of flower (× 1); 4, petal (× 2/3); 5, stigmas pulled apart (× 2); 6, fruit (× 1); 7, seed (× 1). (Reprinted from Exell, 1961, pl. 88, by courtesy of the editorial board, *Flora Zambesiaca*; given there as *Azanza garckeana.*)

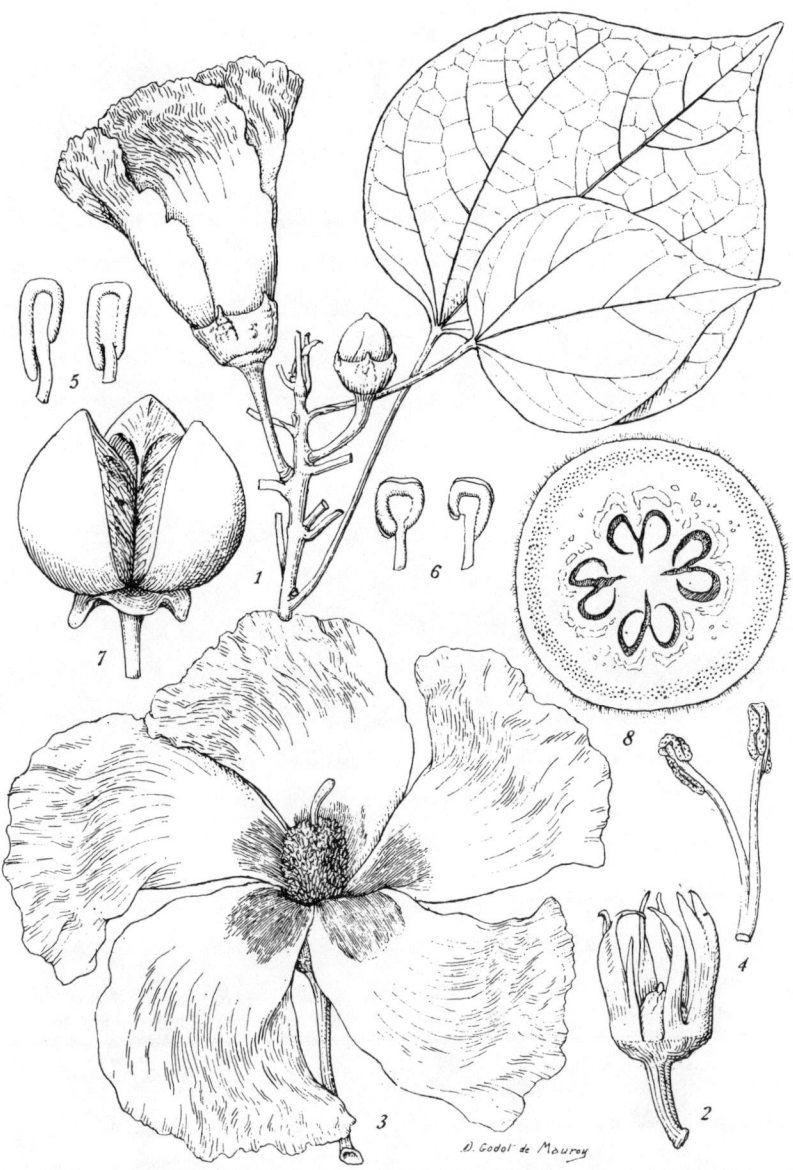

Fig. 43. *Thespesia gummiflua*. 1, flowering branch (× 2/3); 2, flower bud with involucellar bracts (× 2); 3, flower (× 2/3); 4, a pair of stamens (× 3); 5, 6, various forms of anthers (× 6); 7, fruit (× 2/3); 8, cross-section of the ovary at the beginning of fruit development (× 3). (Reprinted from Capuron, 1968, pl. 1.)

Azanza zollingeri Alefeld, *loc. cit.*

Azanza acuminata Alefeld, *op. cit.*, 299.

Abelmoschus zollingeri (Alefeld) C. Mueller in Walpers, Ann. Syst. 7:407. 1868.

Illustrations: Cavanilles, Diss. 3:154, t. 56, fig. 2; Wight, 1838, pl. 5; Blanco, 1879, pl. 355; Hu, 1955, pl. 12; Hochreutiner, 1955, pl. 3, figs. 4–6.

Distribution: India, Indochina, Java, and Borneo to East Africa and the Philippines, where perhaps introduced.

16*b*. *Thespesia lampas* var. *longisepala* Borssum Waalkes, Blumea 14:118. 1966.

Illustration: van Borssum Waalkes, 1966, pl. 14*c*.

Distribution: Borneo.

17. *Thespesia thespesioides* (R. Brown ex Bentham) Fryxell, comb. nov.

Basionym: *Fugosia thespesioides* R. Brown ex Bentham, Fl. Austral. 1:220. 1863.

Fugosia flaviflora F. von Mueller, Fragm. 5:44. 1865.

Gossypium thespesioides (R. Brown ex Bentham) F. von Mueller ex Todaro, Relaz. 103. 1877.

Gossypium flaviflorum (F. von Mueller) F. von Mueller ex Todaro, *op. cit.*, 105.

Hibiscus flaviflorus (F. von Mueller) Kuntze, Rev. Gen. Pl. 1:69. 1891.

Cienfuegosia flaviflora (F. von Mueller) Hochreutiner, Ann. Cons. Jard. Bot. Genève 6:56. 1902.

Cienfuegosia thespesioides (R. Brown ex Bentham) Hochreutiner, *op. cit.*, 58.

Notoxylinon thespesioides (R. Brown ex Bentham) Lewton, J. Wash. Acad. Sci. 5:306. 1915.

Notoxylinon flaviflorum (F. von Mueller) Lewton, *op. cit.*, 307.

Thespesia lampas (Cavanilles) Dalzell ex Dalzell & Gibson var. *thespesioides* (R. Brown ex Bentham) Fryxell, Austral. J. Bot. 13:97. 1965.

Illustrations: Todaro, 1878, pl. 10, figs. 2, 6; Fryxell, 1965*a*, fig. 3.

Distribution: northern Australia.

Excluded Species:

Thespesia altissima (Blume) Spreng. = *Neesia altissima* (Blume) Blume

Thespesia brasiliensis Spreng. = *Luehea divaricata* Martius ?

Thespesia campylosiphon (Turcz.) Rolfe = *Hibiscus campylosiphon* Turcz.

Thespesia rehmannii Szyszylowicz = *Cienfuegosia gerrardii* (Harv. & Sonder) Hochreutiner

Thespesia peekelii (Ulbrich) Borss. = *Cephalohibiscus peekelii* Ulbrich
Thespesia tomentosa Presl = *Hampea tomentosa* (Presl) Standl.
Thespesia thurberi Alefeld = *Gossypium thurberi* Todaro

Since the treatment of *Thespesia* presented here is essentially new, some elaboration is appropriate. As indicated in chapter 1, the species dealt with here have been segregated to several small genera by some workers, notably Urban (1912, 1924), Lewton (1928, 1933), Howard (1949), and Exell and Hillcoat (1954), but have been retained in one genus here, following the inclusive concept of *Thespesia* of Bentham and Hooker (1863), Baker (1897), J. B. Hutchinson (1947a), and van Borssum Waalkes (1966). Kearney (1951) took an intermediate position, recognizing the genera *Montezuma* and *Ulbrichia* but submerging *Atkinsia*, and J. Hutchinson (1967) followed Kearney to some extent, recognizing *Montezuma* and *Thespesiopsis* but submerging the remaining segregate genera in *Thespesia*.

I adopt the inclusive concept of *Thespesia* and cite the following three reasons in support of this position:

1. The ability to hybridize *Thespesia populnea* and *T. garckeana* cited previously (Fryxell, 1968a, p. 301, fn. 5) argues forcefully for a congeneric relationship of these species.
2. The narrow conception of *Thespesia* has placed great weight on the indehiscent capsule of *T. populnea* (the type species) as defining the generic limit of *Thespesia*. The capsule is imperfectly dehiscent, however, in *T. populneoides* (Fig. 34), a species that is so similar to *T. populnea* that the two were commonly confused until Fosberg and Sachet (1972) noted their distinguishing features. This observation considerably weakens the importance of the character of capsule dehiscence in delimiting the genus and thus strengthens the inclusive view.
3. The treatment here of a greater number of species than has been considered by previous authors permits a more comprehensive grasp of the variation encompassed in *Thespesia* sens. lat. and finds a lack of discontinuities in this variation such as are required to establish segregate genera (cf. McVaugh, 1954, pp. 15–17).

The continuous nature of this variation deserves elaboration. There is, of course, a discontinuity at the sectional level separating the groups recognized as section *Thespesia* and section *Lampas*. Although Capuron (1968) discounted this discontinuity and recommended the abandonment of the sectional distinction, he did so from a consideration of only three species and not from a consideration of the variation pattern of the entire genus. One might conceivably argue that these taxa be recognized in generic rank, but the evidence does not seem strong enough at this time to warrant doing so. The fifteen species included in section *Thespesia*, however, present a different problem. At first glance it appeared to me that at least two sections should be recognized here, one centering around *T. populnea* (Fig. 33), and the other around *T. garckeana* (Fig. 42). Several correlated characters seem to support this interpretation: nature of indumentum, capsule dehiscence, number of bracts in the involucel, persistence of the involucel, presence of involucellar nectaries, and simple vs. lobed leaves. But a more careful examination of the entire group of species, especially those more recently described or less well known, shows that these characters are variously recombined in such a way that an essentially continuous series can be proposed connecting the character expressions of the very different species *T. populnea* and *T. garckeana*. Without going into an elaborate exposition of the multivariate variability of these plants, I will simply note such examples as *T. beatensis* recombining the lobed, pubescent leaves of *T. garckeana* with the trimerous involucel of *T. populnea*; *T. fissicalyx* recombining the multipartite involucel of *T. garckeana* with the lepidote indumentum, indehiscent capsules, and simple leaves of *T. populnea*; and *T. gummiflua* recombining the multipartite, deciduous involucel and dehiscent capsule of *T. garckeana* with the lepidote indumentum and simple leaves of *T. populnea*. These (and other) examples reveal a lack of discontinuity in the variation pattern and require the acceptance of a broadly conceived section *Thespesia* without any subdivision into subordinate taxa. The presentation of the key (p. 84) portrays, to some extent, the variation patterns among the species.

Thespesia thespesioides has been elevated to specific rank. When I initially perceived that this plant was in fact a *Thespesia* (Fryxell, 1965a) and made the transfer, I was so struck with its close similarity to *T. lampas* that I emphasized the similarity by retaining it as a variety of *T. lampas*. Greater familiarity with the plant (from having observed living specimens) and a fuller understanding of the genus lead me to realize that these two taxa are sufficiently differentiated to be recognized in specific rank.

One of the striking (perhaps unique) characters of *Thespesia cubensis* is its lack of stipules, a character expression that is very rare in the Malvaceae. It is often difficult to be sure whether or not stipules are present when only dried specimens are available for examination, because the stipules of these plants are often deciduous at a very early stage and are sometimes inconspicuous, leaving scars that are easily overlooked or misinterpreted. I therefore want to make it clear that my statement of the estipulate condition of *T. cubensis* is based on observation of living plants. The stipules are simply not there, even in vestigial condition. The stipules in certain other species of *Thespesia* that are similar in many respects to *T. cubensis*, especially of *T. mossambicensis* and *T. acutiloba*, are therefore of interest. In both cases the stipules are very small (1 mm) and deciduous very early. This may indicate an affinity of the three species, in spite of their geographic separation. More strikingly, the distinctive and well-developed domatia characteristic of *T. cubensis* (each with a prominent tooth or projection) are also found in *T. acutiloba*, although domatia are essentially lacking in *T. mossambicensis*.

3

Comparative Morphology

*The celebrated DeCandolle, however, when enumerating the
several species [of Gossypium], stated, that all were uncertain,
and that no genus required more the labours of a monographist
who could describe them from living specimens.*
John Forbes Royle, 1851

It is better to cite DeCandolle than to curse the darkness.
Richard Eyde, 1968
(with presumed apologies to
the Christopher Society)

THE following discussion will place the emphasis on the comparative
aspects of morphology rather than attempting to be a definitive treatise
on the detailed morphology of the tribe. The emphasis will be on the
range of variation of significant morphological characteristics and on
discontinuities in this variability, as well as on morphological patterns
common to all members of the tribe. This approach is in keeping with
the central thrust of the work as a whole.

The morphological traits discussed in this chapter are a principal
basis for the classification presented in the previous chapter and for the
discussions of ecology, phytogeography, and evolution that occupy following chapters.

The Embryo

THE embryo is an appropriate subject with which to begin because it is
the organ that has all other organs inherent in it and because its structure is very distinctive in the tribe Gossypieae. In fact, the distinctiveness of the embryo is one of the principal bases for the delimitation of
the tribe (Fryxell, 1968a).

According to the studies of Martin (1946) on the comparative
morphology of angiosperm embryos, the embryos of the tribe Gossypieae are among the most fully developed and complex of any angiosperms (Figs. 44, 45). The cotyledons are complexly folded in a way

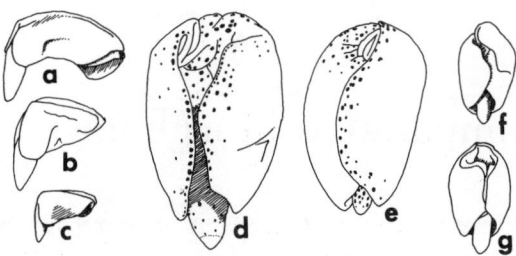

Fig. 44. Comparative illustrations of embryos of *Hibiscus* and of genera of the Gossypieae: a, *Hibiscus meraukensis* Hochr.; b, *Hibiscus cannabinus* L.; c, *Hibiscus caesius* Garcke; d, *Thespesia populnea;* e, *Hampea appendiculata;* f, *Cienfuegosia hearnii.* (Redrawn from Fryxell, 1968a.)

that completely encloses the epicotyl and hypocotyl so that, when extracted from the seed coat, the embryo generally has only the tip end of the hypocotyl visible. This contrasts with the embryos found elsewhere in the Malvaceae, especially in the tribe Hibisceae, where the hypocotyl is generally not enclosed by the cotyledons even though the cotyledons are large and well developed and may be folded back upon themselves. The only exception to this pattern is *Cephalohibiscus peekelii*, which has a relatively simpler embryo, such as that found in *Hibiscus* spp., but which in all other respects belongs in the Gossypieae.

In most angiosperms a reciprocal relationship exists between the relative development of the embryo and the endosperm in the mature seed. This reciprocity necessarily results from the developmental relationship between them: the endosperm is the tissue that nurtures the developing embryo during seed maturation; it may be viewed as the nutritive substrate for the growth of the embryo. Seeds with relatively simple, undifferentiated embryos often have these embryos immersed in a copious endosperm. Seeds with complex, well-developed embryos, on the other hand, often have little or no endosperm remaining in the mature seed, it having been exhausted in nurturing the development of the more complex embryo. In one sense, these types may be said to differ only in the timing of the period of embryo dormancy, that is, the stage of development that we call the mature seed. The endosperm will be used up to support the embryo's development; it is simply a question of whether this exhaustion occurs during seed development and before seed maturation or during germination and after seed maturation.

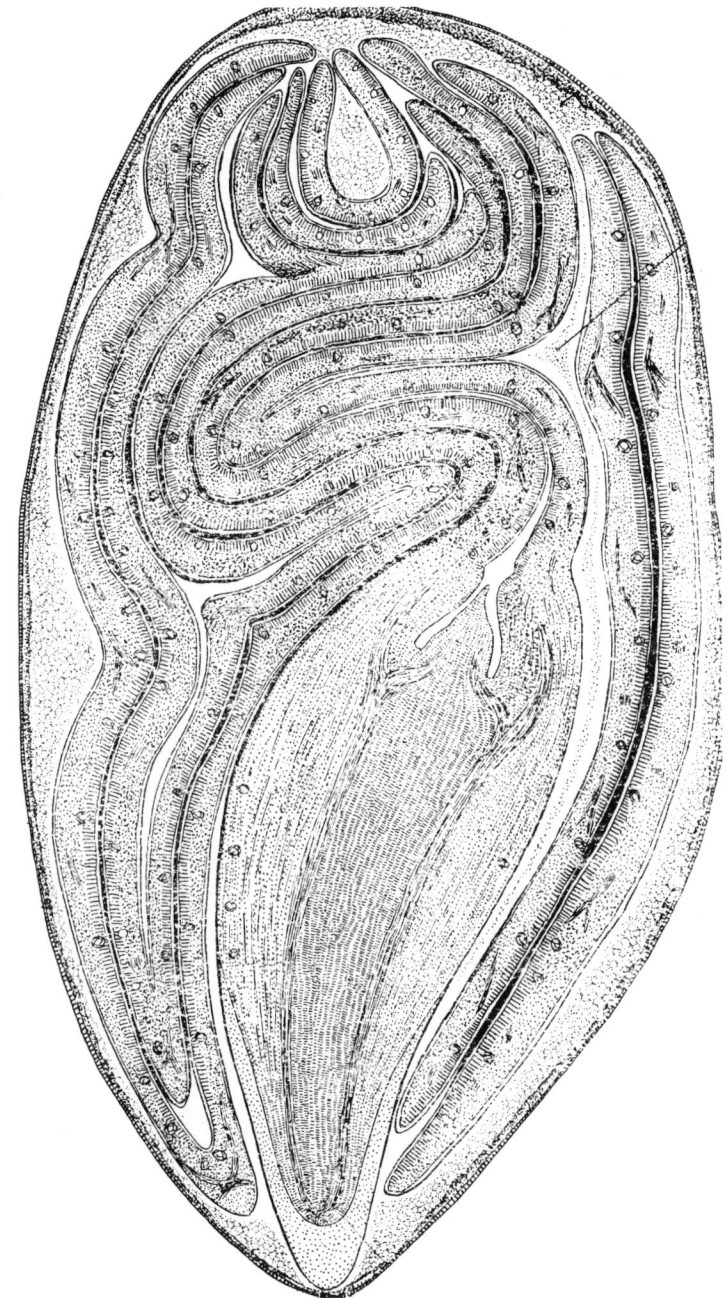

Fig. 45. Cross-section of seed of *Gossypium*. (Reprinted from P. A. Baranov and A. M. Maltzev, *The Structure and Development of the Cotton Plant.* Moscow-Leningrad: Ogis-Isogis, 1937, pl. 42, fig. 7.)

The extreme case is found in the tribe Gossypieae. As already stated, but with the exception of *Cephalohibiscus*, the embryos are large and complex in their development. By the time the seed is matured, only remnants of endosperm tissue remain, folded among the lobes of the cotyledons and surrounding the hypocotyl. These remnants are perhaps most noticeable in *Thespesia*, but even there the role of the endosperm as a food-storage organ of the seed is insignificant. In the Gossypieae, food is stored in the seed largely in the form of oil droplets in the cells of the cotyledons. (It is this oil that is extracted commercially from cottonseed and finds its way to our dinner table as margarine. The embryo is rich in oil and protein but has essentially no starch.) More commonly (as in *Hampea* or *Gossypium*) the endosperm degenerates to a thin membrane that encloses the embryo at maturity, much as the amnion encloses the mammalian newborn. The endosperm is a negligible structure in the mature seed.

The epicotyl in the mature embryo is only an undifferentiated meristem, completely enclosed by the cotyledons, that begins its development only after germination.

The Seedling

UPON germination the hypocotyl elongates and the cotyledons unfold. The former is negatively, the latter is positively, phototropic. As a result, the hypocotyl becomes an aggressive taproot, and the cotyledons quickly turn green and spread themselves in the sunlight as functioning photosynthetic organs almost from the beginning of germination. Only after the seedling is established to this extent, with a functional root system (probably branching by this time) and with photosynthetically active cotyledons, does the epicotyl begin to expand and develop the first true leaves.

The form of the expanded cotyledons is characteristic of the tribe, but interesting variations occur within the tribe. Typically the cotyledons are transversely oblong-elliptic (or sub-reniform) in shape, with a truncate apex, a somewhat cuneate base, and a relatively short stalk (Fig. 19*K*).

In some species the apex of the cotyledons is more or less emarginate, so that the cotyledons are obcordate or have a bilobed aspect. This characteristic is weakly expressed in certain species of *Cienfuegosia* (for example, *C. argentina*) and is more strongly expressed in *Gossypium* (*G. sturtianum* and *G. robinsonii*).

Another more specialized characteristic occurs in at least one species of *Hampea* (*H. nutricia*) but not others (for example, *H. rovirosae*) and in a few species of *Gossypium* (such as *G. pilosum* and *G. populifolium*) but not most of them. In these species the individual cotyledon is more or less bilobed and asymmetrical. The lobe that is external as the cotyledons are folded in the embryo is the larger, and the lobe that is internal is the smaller. Upon germination of the seed, the cotyledons open up and flatten out in the normal way but display their lop-sided shape. The two larger lobes are not adjacent to one another but are diagonally opposite, as are the two smaller lobes. The larger lobes are more or less succulent, presumably containing extra food reserves. It is difficult to speculate what might be the adaptational advantages of this specialization.

The cotyledons have the capability to persist for an indefinite period. After one or more true leaves have developed, and if the young plant is subjected to stress, abscission layers will quickly form and the cotyledons will fall. But if conditions remain favorable for the growing plant, the cotyledons may persist and remain green and apparently functional even after the main stem begins secondary thickening.

The Stem

THE nature of the stem determines the growth habit of the plant. In the Gossypieae, growth habits range from procumbent herbs (such as *Cienfuegosia sulfurea*) to large trees (such as *Thespesia cubensis*) and span a variety of intermediate patterns.

All members of the tribe are perennial, and most are woody. The arborescent form is most fully developed in *Cephalohibiscus peekelii*, trees of which reach 20–30 m in height, and in several species of *Thespesia* (for example, *T. patellifera*, *T. robusta*, *T. fissicalyx*, and *T. cubensis*) and of *Hampea* (for example, *H. albipetala*, *H. thespesioides*, and *H. appendiculata*), all of which attain comparable stature. These are among the larger representatives of the Malvaceae. Such dimensions are equaled or exceeded within the family only in *Hibiscus*, especially the New Guinea species such as *H. archboldianus* Borss. and the pantropical *H. tiliaceus* L., and in *Robinsonella*, especially the Mexican *R. mirandae* Gómez-Pompa.

Trees of more modest dimensions are characteristic of *Kokia*, *Lebronnecia*, most of the species of *Hampea*, many of those of *Thespesia*, and of *Gossypium* sect. *Erioxylum*. Large shrubs up to several meters tall also occur in *Thespesia*, *Hampea*, and some species of *Gossypium*, the distinction between a small tree and a large shrub being partly a matter of circumstance and partly a matter of the degree to which the plant develops a single dominant trunk instead of a multistemmed pattern of growth. Relatively small shrubs less than 2 m tall occur in other species of *Gossypium*, in *Gossypioides*, and in a few species of *Cienfuegosia*. Herbaceous or suffruticose growth patterns are typical of a few species of *Gossypium* (for example, *G. triphyllum* and *G. stocksii*) and most species of *Cienfuegosia*.

The greatest diversity of growth habit is found within the genera *Gossypium* and *Cienfuegosia*. In *Gossypium* we find arborescent plants such as *G. laxum*, large shrubs such as *G. trilobum*, small shrubs such as *G. sturtianum*, subshrubs such as *G. triphyllum*, scandent plants such as *G. longicalyx*, and more or less procumbent plants such as *G. stocksii* and *G. populifolium*. In *Cienfuegosia* the range includes small- to moderate-sized shrubs (for example, *C. gerrardii* and *C. affinis*), subshrubs (for example, *C. rosei* and *C. somaliana*), erect, more or less herbaceous plants (for example, *C. yucatanensis*), and several trailing or procumbent herbs (such as *C. sulfurea* and *C. argentina*). Even the fully herbaceous plants have perennial rootstocks.

There is little to distinguish the stems in gross morphological terms except to note that in *Gossypioides* and in two species of *Gossypium* (*G. thurberi* and *G. trilobum*) the stems are pentagonal in cross-section and almost have an appearance of being winged. This same character is weakly developed in several species of *Cienfuegosia*, most markedly in *C. tripartita* and *C. hitchcockii*.

Leaves

THE leaves of the Malvaceae generally, and of the Gossypieae in particular show characteristic patterns (Fig. 46). Basically, the leaves are spirally disposed, stipulate, petiolate, palmately nerved, cordate in shape, and (in the Gossypieae) usually provided with foliar nectaries on the undersurface of the leaf on one or more of the principal nerves.

Exceptions occur to most of these characteristics. In those species with procumbent stems (such as *Cienfuegosia argentina*), the leaves are distichous rather than spirally disposed, and they are also distichous in the arborescent *Gossypium lobatum* (Fig. 23). Stipules do not occur in *Thespesia cubensis* although they are characteristic otherwise of the family (even though they are vestigial in the genus *Batesimalva*).

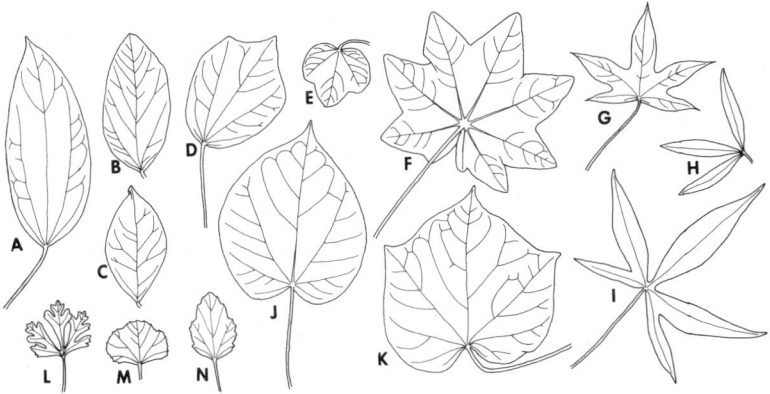

Fig. 46. Leaf outlines, illustrating variations in form found among members of the Gossypieae: A, *Hampea longipes;* B, *Cienfuegosia affinis;* C, *Gossypium cunninghamii;* D, *Hampea trilobata;* E, *Gossypium somalense;* F, *Kokia cookei;* G, *Gossypium robinsonii;* H, *Gossypium triphyllum;* I, *Gossypium thurberi;* J, *Lebronnecia kokioides;* K, *Hampea mexicana;* L, *Cienfuegosia digitata;* M, *Cienfuegosia hildebrandtii;* and N, *Cienfuegosia drummondii.*

Two species (*Cienfuegosia affinis, Gossypium cunninghamii* [Fig. 46B, C]) have lanceolate leaves with pinnate venation rather than broader, palmately nerved leaves. And finally, the species of *Cienfuegosia* subgen. *Cienfuegosia* lack foliar nectaries, as do occasional species elsewhere in the tribe, such as *Gossypium tomentosum* and *G. gossypioides.* All of these exceptions may be regarded as specializations of the more common, basic pattern.

The least specialized form of leaf is undoubtedly the simple, entire, cordate leaf that is characteristic of most of the species of *Thespesia,* of *Lebronnecia* (Fig. 46J), of many species of *Hampea,* and of *Gossypium.* One common variant of this form is the broader leaf with a moderate development of leaf lobing, as occurs in some species of *Thespesia* (for example, *T. garckeana* and *T. trilobata*), *Hampea* (for example, *H. mexicana* [Fig. 46K] and *H. platanifolia*), and *Gossypium* (for example, *G. davidsonii* and *G. laxum*), and in *Gossypioides* spp. and *Cephalohibiscus peekelii.* A distinctive type of leaf lobing is found in the species of *Kokia* (Figs. 31, 32, 46F).

Another common variant of the "basic" cordate leaf is the narrower, elliptic form, such as that found in several species of *Hampea* (as in *H. longipes* [Figs. 29, 46A], *H. micrantha,* and *H. integerrima*) and in a few other species (for example, *Cienfuegosia affinis* [Figs. 10, 46B], *C. yucatanensis,* and *Gossypium cunninghamii* [Fig. 46C]). This form usually is accompanied by a leaf base that is truncate to cuneate instead of cordate.

More complex types of leaf lobing occur in certain species of *Gossypium* and *Cienfuegosia*. Moderately incised leaves are typical of such species as *Cienfuegosia rosei* and *Gossypium stocksii* (Fig. 26); more deeply lobed leaves are found in such species as *Gossypium thurberi* (Figs. 21, 46*I*) and *G. robinsonii* (Figs. 18, 46*G*); trifoliolate leaves occur in *Gossypium triphyllum* (Figs. 19, 46*H*) and *Cienfuegosia sub-ternata*; and secondary lobing occurs in *Cienfuegosia digitata* (Fig. 46*L*), *C. hasslerana,* and *Gossypium capitis-viridis.*

Leaves have been reduced to a flabelliform shape in *Cienfuegosia* sect. *Garckea* (Figs. 7, 46*M*). They are highly variable in form in *Cien-fuegosia heterophylla* (Fig. 9) and *C. tripartita.* Distinctive basal auri-cles occur in *Gossypium sturtianum* var. *nandewarense* and in several species of *Hampea* sect. *Trianchonia.*

Dentate or serrate (or sometimes crenate) leaf margins are found in many species of *Cienfuegosia* (Figs. 11, 12, 46*M, N*) but in none of the other genera of the tribe. Both African and American representa-tives of *Cienfuegosia* have serrate-margined leaves.

The petioles in the tribe are all pretty much alike. They may be longer in some species and shorter in others and have differing kinds of vestiture, but there are few characters to distinguish them. About the only distinguishing feature occurs in the two species of *Gos-sypioides*, which have markedly quadrangular petioles, a characteristic that seems to be related to the pentangular stems. However, the same quadrangular petioles and pentangular stems may be found in *Gos-sypium trilobum* (and more weakly expressed in *G. thurberi*). Several species of *Cienfuegosia* have canaliculate petioles.

It has already been noted that stipules are absent in *Thespesia cubensis.* Apart from this exceptional instance, however, stipules are characteristic of the tribe and are of three types. Most commonly stipules are subulate or linear; these may be long or short, prominent or inconspicuous, symmetrical or somewhat falcate. They are rigid and persistent (even after leaf abscission) in *Cienfuegosia humbertiana.* Less commonly, in *Gossypioides kirkii* (Fig. 14), *Cienfuegosia hilde-brandtii,* and *Cienfuegosia* sect. *Friesia* (Fig. 13), the stipules are broad and foliaceous, auriculate-clasping, markedly asymmetrical, and usu-ally prominent. The third type occurs only in *Cienfuegosia gerrardii.* The stipules of this species are symmetrically oblong or elliptic.

broadly foliaceous, and cuneate-sessile; they are very prominent and unlike anything else in the tribe.

The foliar nectaries that occur on the undersurface of the leaf are not unique to the tribe but are certainly characteristic of it. Similar structures occur elsewhere in the Malvaceae (as in *Urena* spp., *Decaschistia* spp., and many *Hibiscus* spp., but never in the tribe Malveae) and in other families as well, and unique stipular nectaries occur in the genus *Radyera* (Fryxell and Hashmi, 1971), but in the Gossypieae the absence of foliar nectaries is the noteworthy characteristic, and such absence must therefore be regarded as a specialization. Foliar nectaries are lacking in the species of *Cienfuegosia* subgen. *Cienfuegosia* and in two species of *Gossypium* (*G. gossypioides* and *G. tomentosum*).

Interesting variations in the form and distribution of these nectaries occur within the tribe. Relatively simple nectaries are found in species like *Lebronnecia kokioides* and *Thespesia populnea* that are little more than an aggregation of exposed secretory cells with little or no elaboration of "gland" structure and bordering protective tissue. This secretory zone near the base of the midrib is often relatively elongated. In *Thespesia danis* this secretory surface occurs on the underside of the petiole rather than on the leaf blade itself. More commonly the nectaries are sunk into the midrib and form a more complex structure (Fig. 47). It is often elongated, with the proximal end of the nectary truncate and the distal end acute, but in other cases the nectary is foreshortened to a nearly circular form.

The number of nectaries per leaf and the position of the nectaries are correlated. Species with only a single nectary (such as *Hampea tomentosa* [Fig. 47], *H. trilobata*, and *Gossypium thurberi*) generally have it located at or near the base of the lamina near the juncture with the petiole. Species with several nectaries (for example, *Gossypioides kirkii*, *Hampea platanifolia*, and *Gossypium robinsonii* [Fig. 18]) have them located one on each of the principal nerves in a relatively more distal position out near (though never at) the apices of the leaf lobes. Intermediate situations are common, with the central nectary on the midrib located near midlamina and with two or more lateral nectaries on lateral primary nerves in a more nearly basal position.

In several species of *Gossypium*, especially the Australian species *G. robinsonii*, *G. australe*, and *G. bickii*, the secretory tissue of the foliar nectaries is colored a bright red, and the nectaries therefore stand out prominently. The involucral nectaries of these species have the same coloration. Red involucral nectaries are rarely found also in *Gossypium barbadense*, but this characteristic does not occur elsewhere in the tribe.

Domatia are characteristic of several species of the tribe, primarily in the genera *Thespesia* and *Hampea* (Fig. 47). These structures occur on the underside of the leaf in the axils, where principal nerves come together at the base of the lamina. A domatium is a flap of tissue that bridges over the more or less triangular area at the juncture of two veins and forms a small "cave" or "house"—hence the name, *domatium*. Since these spaces are sometimes used as a refuge or protection by mites (acarids), they are sometimes called acarodomatia.

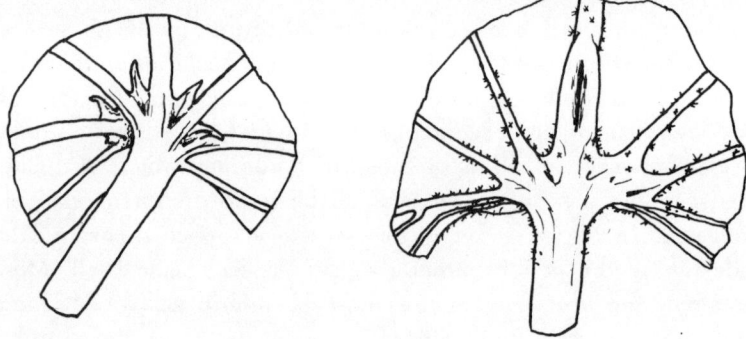

Fig. 47. Domatia, as found in *Thespesia cubensis* (left) and *Hampea tomentosa* (right).

Mites, however, are generally detrimental to the plants they infest, so it is difficult to conclude that these domatia have any adaptive value to the plant in serving as protection for the mites, although such hypotheses have been advanced. Perhaps detailed study of the relationships between mites and their host plants will reveal some unsuspected symbiosis, but a review of the question by Jacobs (1966) does not present any positive leads.

Perhaps the most fully developed domatia occur in *Thespesia cubensis* (Fig. 47), in which a toothlike callosity or projection is found. Similar domatia are found in *T. acutiloba*. They are less well developed in *T. grandiflora*, *T. populneoides*, and *T. populnea* and are absent in such species as *T. lampas*, *T. beatensis*, *T. danis*, and *T. garckeana*. In *Hampea*, domatia occur in *H. latifolia*, *H. longipes*, and *H. tomentosa* (Fig. 47), and they sometimes occur in *H. integerrima*; they are absent in other species.

Jacobs (1966), in a thorough review of the subject of domatia, notes that they occur only on ligneous dicots and that they seem to occur only on plants of humid habitats. The data reviewed here for the tribe Gossypieae generally conform to these restrictions. Domatia in this tribe are confined to the arborescent genera *Thespesia* and *Hampea* and within these genera occur on those species from relatively more mesic habitats. Two cases are worth specific comment in this connection, however. One instance of a species with well-developed domatia is *Hampea tomentosa*, which occurs in western Mexico (in Colima and adjacent states), where it is subjected to intense seasonal aridity. This example apparently contradicts Jacobs' view that domatia do not occur under arid conditions; however, at the time that growth occurs and the leaves are developed (that is, during and following the rainy season), conditions are certainly *not* arid, so the generalization seems to hold up, even in this case.

A second case concerns the littoral species *Thespesia populnea* and *T. populneoides*. The habitats they occupy are varied in terms of amount of incident rainfall, but if one equates environmental salinity with physiological moisture stress, they may be said to occur in arid habitats. Their possession of domatia thus seems to be a special case.

Inflorescence

THE disposition of flowers in this tribe follows three basic patterns. The simplest pattern involves isolated flowers borne singly in the axils of the leaves. The peduncles bearing these flowers may be long or short, articulated or not, and sometimes bracteate at the point of articulation. This pattern of the solitary disposition of the flowers occurs in most species of *Thespesia* (as in *T. populnea* [Fig. 33]), in *Kokia* (Figs. 31, 32), in *Cienfuegosia* subgen. *Cienfuegosia* (Figs. 10–13), in *Gossypium* sect. *Erioxylum* (Figs. 22, 23), exceptionally elsewhere in *Gossypium*, and in a few species of *Hampea*.

The second pattern of inflorescence development involves the production of lateral flowering branches, with the individual flowers arranged in a sympodial pattern. These axillary sympodial branches may bear many flowers, one at each node, or they may be reduced under adverse growing conditions to a single axillary flower. The sympodial inflorescence is characteristic of *Gossypium* (Fig. 20), *Gossypioides* (Fig. 14), *Cienfuegosia* subgenus *Articulata*, *Lebronnecia*, and *Cephalohibiscus* (Fig. 6). In *Cephalohibscus* the inflorescences are typically reduced to two-flowered, seemingly umbellate structures, but this pattern is believed to be simply a modification of the sympodial pattern that occurs widely in the tribe.

The third basic pattern, axillary fascicles of flowers, is primarily found in the genus *Hampea* (Figs. 29, 30) but also occurs as an isolated specialization in *Gossypium lobatum*. In *Hampea* the flowers are typically borne on short axillary pedicels (although these may be 5–6 cm long in *H. longipes* or even 11 cm long in *H. breedlovei*) that are in clusters of as many as a dozen or more flowers per axil. The number of flowers per cluster varies from one species to another, sometimes (as in *H. latifolia* or *H. rovirosae*) being only one or two. In the dioecious species of *Hampea*, the number of flowers per axillary fascicle is usually greater on the staminate trees than it is on the pistillate trees.

A striking exception to this pattern in *Hampea* is *H. micrantha*. In this species the flowers, instead of being fasciculate, are borne on short, few-flowered but branched axillary peduncles. Such branched peduncles do not occur elsewhere in the genus. Superficially the inflorescences of *H. micrantha* do not appear to differ significantly from the fascicles of the other species, but close inspection reveals the difference.

Finally, one exception to the overall pattern for the tribe should be noted. *Cienfuegosia heteroclada* (Fig. 8) has a specialized growth habit that appears to be an adaptation to periodic burning of the grasslands where it is native. It is nearly acaulescent, flowering and fruiting directly from the perennial rootstock. The flowering shoots arise before the leaves. The flowers are borne on short (1–3 cm), leafless racemes with minute bracts at the nodes. This inflorescence structure, like the dioecious flowering pattern of *C. heteroclada*, differs markedly from the other species of *Cienfuegosia*—indeed, from the remainder of the tribe.

The Pedicel and Involucel

REGARDLESS of the simplicity or complexity of the inflorescence, the individual flower of all the species of the tribe is borne on an individual pedicel, and that pedicel (with certain exceptions) is surmounted by an involucel.

The pedicels themselves present few variations of interest. They may be relatively longer or shorter, more or less pubescent, more or less angular in cross-section, and more or less slender. Some, especially among the larger-flowered species, are notably enlarged toward the apical end (Fig. 48B). But beyond these variations there is little to distinguish them.

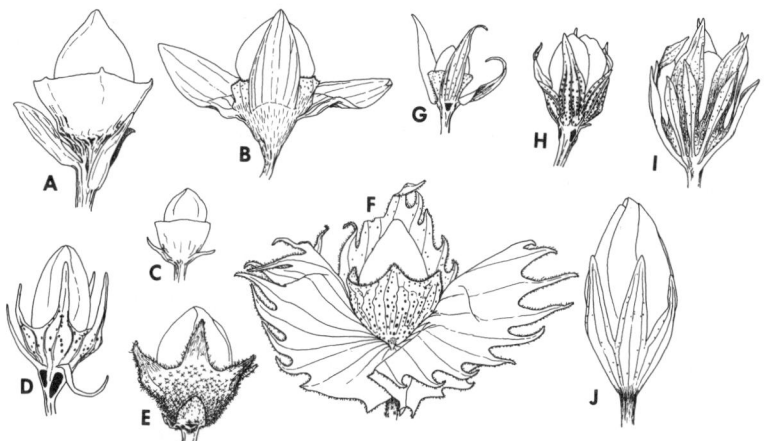

Fig. 48. Buds, illustrating variations of calyx and involucel among members of the Gossypieae: A, *Thespesia populnea;* B, *Thespesia danis;* C, *Hampea integerrima;* D, *Gossypium australe;* E, *Gossypium lobatum;* F, *Gossypium hirsutum;* G, *Gossypium thurberi;* H, *Cienfuegosia yucatanensis;* I, *Cienfuegosia drummondii;* and J, *Cienfuegosia ulmifolia.*

When we examine the involucel, however, and the nectaries that are an integral part of it, we find a wide range of morphological variations (Fig. 48). The tribe Gossypieae is distinctive both for the presence and for the trimerous nature of these nectaries. Where similar nectaries do occur elsewhere (as in certain Bombacaceae), they are generally of an indefinite number. Several species of *Hibiscus* sect. *Furcaria* have equally distinctive nectaries on the calyx, but these are pentamerous in nature, are borne on the midribs of the calyx lobes, and are not homologous to the nectaries of the involucel of the Gossypieae. Nectaries also occur on the *tips* of the calyx lobes of *Hibiscus calyphyllus*, but again, these structures are not homologous to those under discussion here.

The nectaries of the involucel as found in the Gossypieae occur at or near the apex of the pedicel, immediately below the insertion of the involucel. The nectaries are always trimerous. They are usually in a whorl, like the bracts of the involucel themselves, but in a few species (*Lebronnecia kokioides, Thespesia beatensis, Thespesia populnea* [Fig. 48A], and *Hampea rovirosae*) both the nectaries (if present) and the bracts are more or less irregularly inserted in what is evidently a relictual spiral arrangement.

The nectaries are often obtriangular in form and sometimes have prominent raised borders (as in *Hampea tomentosa*), and, in a few Australian species of *Gossypium* mentioned above, the nectaries of the involucel, in concert with the foliar nectaries, may be colored a bright red. These nectaries are capable of exuding copious amounts of nectar, although the quantity is dependent upon the species and the conditions under which the plant is growing.

In the domesticated species of *Gossypium*, an additional set of nectaries occurs in association with the involucel. The trimerous nectaries that are characteristic of the tribe occur immediately below each of the three bracts that subtend the flower. They thus subtend the bracts. The extra set of three nectaries occurs alternate with and slightly above the three bracts. The subtending nectaries are often called the "outer involucellar nectaries," and the additional ones are called the "inner involucellar nectaries." The latter set is actually inserted on the base of the calyx (Fig. 48F). In certain genotypes of these variable cultigens (especially in *Gossypium arboreum*) the nectaries may be suppressed, and plants are known in which the "outer" nectaries are absent and the "inner" ones present, an expression not otherwise known in the tribe. More commonly it is the "inner" set that is suppressed.

The involucel shows several distinct variations. In some species of *Thespesia* (Figs. 42, 43), many species of *Cienfuegosia* (Fig. 48H, I), and two species of *Hampea* (*H. tomentosa* and *H. rovirosae*) the number of bracts of the involucel is greater than three (commonly nine but occasionally fewer, evidently by abortion), and it may be up to a large and indefinite number (about twenty) in *T. multibracteata* (Fig. 39). These bracts can often be shown to have a trimerous nature, with nine bracts in groups of three, each group surmounting an individual nectary and each group often having the central bractlet slightly larger (as in *Cienfuegosia hildebrandtii*, *Thespesia lampas*, and *Hampea tomentosa*). In other cases (for example, *Cienfuegosia heterophylla* [Fig. 9]) the bracts are more evenly and regularly distributed around the apex of the pedicel and are of a uniform size, and the trimerous nature of the involucel is shown only by the trimerous nature of the nectaries. In *Cienfuegosia* sect. *Robusta* the nectaries are lacking and the involucel is of eight to ten bractlets so that its trimerous nature is completely masked. In cases like *Thespesia multibracteata* the trimerous nature of the involucel is obscured by its very complexity. The involucel of these species is usually persistent, but it may be deciduous, as, for example, in *Thespesia garckeana* and *Hampea tomentosa*.

Some of those species with the number of bractlets of the involucel approximating nine show a variable degree of lateral fusion of adjacent bractlets. This lateral fusion is generally among the three bracts associated with a common nectary and may be considered transitional to the fully trimerous involucel that is common in the tribe; *Cienfuegosia hildebrandtii* is a species showing a slight degree of lateral fusion, and in *C. heteroclada* the fusion is so fully expressed that its involucel *is* trimerous, but the individual bracts are sometimes apically laciniate or even two- or three-parted, suggesting a connation of parts.

In some species of *Thespesia*, most species of *Hampea*, and all species of *Gossypium, Gossypioides, Kokia, Cephalohibiscus*, and *Lebronnecia* the bracts of the involucel are basically three in number. They are of variable form; the entire gamut of form is to be found in *Gossypium*.

In some species (for example, *Gossypium bickii, Cephalohibiscus peekelii, Lebronnecia kokioides,* and most species of *Hampea*) the bracts of the involucel are narrowly linear, inconspicuous, and sometimes caducous. In other cases they are linear but broader (as in *Thespesia populnea* and *Gossypium anomalum*) and somewhat foliaceous. In *Gossypium* subsect. *Erioxylum* (Fig. 48E) the involucel is reduced to three small, persistent scales that are about as broad as long.

In *Kokia* (Figs. 31, 32), *Gossypioides* (Figs. 14, 15), and many species of *Gossypium* (Fig. 48F) the bracts of the involucel are highly developed into broad, foliar, usually cordate structures. These involucels are very prominent and characteristic. They usually enclose (and protect?) the bud. In fact, in *Gossypium gossypioides* (Fig. 24) the margins of the bracts in the bud are joined by interlocking hairs and are forced apart only at anthesis when the rapidly expanding corolla pushes its way through and separates the bracts. In some species the bracts also more or less enclose the mature fruits. The bracts may be entire (as in *Kokia* spp. [Figs. 31, 32], *Gossypium gossypioides* [Fig. 24], *G. sturtianum* [Fig. 16], and *G. longicalyx* [Fig. 28]); they may be apically dentate (as in *Gossypium capitis-viridis* and *G. sturtianum* var. *nandewarense* [Fig. 17]); or they may be complexly laciniate (as in *Gossypium raimondii* and *G. davidsonii*). In *Gossypium armourianum, G.*

harknessii, and *G. turneri* the involucel is deciduous, but otherwise in *Gossypium,* and in other species with foliaceous bracts, the involucel is persistent. In fact, in some species (for example, in *G. gossypioides* and *G. somalense*) the involucel is an integral part of the dispersal unit, falling intact with the mature fruit by abscission at the base of the pedicel. Presumably the bracts aid in dispersal by tumbling in the wind (cf. chapter 4).

In the species of *Cienfuegosia* sect. *Friesia* (Fig. 48*J*) the involucel is completely suppressed.

The Flower

THE flowers of the Gossypieae range from those less than 2 cm in diameter (*Hampea sphaerocarpa* and *H. stipitata*) to those that are showpieces 15 cm in diameter (*Thespesia grandiflora, T. beatensis,* and *Kokia drynarioides*). *K. kauaiensis* (Fig. 32) is described as having flowers over 20 cm in diameter. These showy-flowered shrubs and trees are deserving of wider consideration in the tropics for planting as specimen ornamentals. Only *T. grandiflora* (Fig. 38) has received much attention in this respect.

The calyx in the Gossypieae (Fig. 48) is uniformly pentamerous and gamophyllous. In those species with truncate calyces, the pentamerous nature is often obscured, but it can usually be demonstrated in the venation, in the presence of five vestigial teeth, or in similar characters.

Calyx lobes are long and clearly distinguished in *Cienfuegosia* (especially in the American species) and in certain species of *Gossypium* (*G. longicalyx* [Fig. 28], *G. lobatum* [Figs. 23, 48*E*], and most Australian species [Fig. 48*D*]). In these well-developed and more deeply lobed calyces, the gossypol glands are often particularly well developed, especially in *Cienfuegosia* sect. *Cienfuegosia* (Fig. 48*H*), but they may also be very obscure, as in *Cienfuegosia* sect. *Friesia* (Fig. 48*J*). In some cases (as in *Gossypium australe*) they are obscured by the indumentum. Ordinarily they are no more prominent in the calyx than in the foliage.

Instead of being clearly five-lobed, the calyces of the tribe are more commonly five-toothed, undulate-margined, or fully truncate. In *Gossypium* the full range of variation is found, from fully truncate forms (*G. thurberi* [Fig. 48*G*] and *G. armourianum*), through undulate-margined species (*G. somalense* and *G. sturtianum*) and short-lobed species (*G.*

incanum and *G. lobatum*), to species such as *G. cunninghamii* with large, deeply lobed calyces. A unique calyx form occurs in *G. trilobum*, in which the calyx has ten irregularly developed caudate projections arising from the margin. The exact homology of the calyx including these structures to the pentamerous calyces of other species of the genus is not fully clear.

In *Lebronnecia* the calyces are five-toothed and clearly pentamerous in their venation. Calyces are fully truncate and chalice-shaped in *Cephalohibiscus* (Fig. 6), in most species of *Thespesia*, and in *Hampea* (Fig. 48C), in which vestiges of the pentamerous condition are evident in a few species, notably *H. platanifolia*. The calyces of *Gossypioides* are nearly truncate, as are those of *Kokia*.

In *Thespesia grandiflora* the calyx is circumscissile at the base and deciduous after flowering. This character is unique to this species and has been used as a basis for segregating it in generic rank, but it seems to me to be better accommodated in *Thespesia*. The calyx lobes of *Kokia* have also been described as abscising, but I am unfamiliar with the sequence of events of floral development, and the calyx in this genus may well be truncate by nature rather than by abscission. Observation of living material will be needed to settle this point.

The corolla in the Gossypieae is uniformly composed of five petals inserted at the base of the staminal column. The petals are distinct and imbricate in their aestivation. They are generally obovate and narrowed at the base to form a more or less distinct claw that is usually pubescent on the margin. The petals of *Cephalohibiscus* are very curiously contorted (Fig. 6). The corolla is presented in a variety of ways, no doubt dependent on relationships to particular pollinators. In some species (*Gossypium aridum* [Fig. 22], *G. areysianum* [Fig. 27], and *Cienfuegosia hearnii* [Fig. 7]) the corolla is narrowly funnelform, scarcely opening much more than enough to admit a large insect to serve as pollen vector. Such corollas, interestingly, are often lavender or purple. More commonly the corolla is broadly funnelform or campanulate. The corolla in such flowers is commonly yellow or white, though sometimes lavender or mauve, and the petals often have a dark (red or maroon) spot on the claw, giving the flower a dark throat. The size of this spot may be quite variable, from covering more than half the petal to vestigial (to completely absent). Flowers of this type characterize most species of *Gossypium*, *Cienfuegosia*, and *Thespesia*.

In *Cienfuegosia yucatanensis* the corolla is fully rotate in form.

This character is unusual for the tribe, being approximated only in *Gossypium thurberi*. These two species have the smallest flowers of their respective genera and are characterized by the reduction of the petal spot (absent in *C. yucatanensis*, vestigial in *G. thurberi*).

The flowers of *Hampea* stand apart from those of the remainder of the tribe in having corollas that are relatively small, more or less white, and lacking a dark throat, with reflexed petals and a distinctive fragrance. They stand apart in other characters, too, such as their disposition in axillary fascicles and their usually unisexual nature. The dioecious condition that is characteristic of most species of *Hampea* (Fig. 49) is in striking contrast to the perfect-flowered species that are characteristic of the balance of the tribe. Only *Cienfuegosia heteroclada* exhibits a similar kind of floral dimorphism, but its exact nature is not yet clearly understood. In *C. heteroclada* a proportion of 90 percent staminate to 10 percent pistillate plants has been reported, whereas a 50–50 sex ratio appears to characterize *Hampea* spp. In neither case has the genetic control of dioecy been analyzed. Elsewhere in the Malvaceae, dioecy is known only in *Kydia* spp. (of uncertain affinities), in the North American *Napaea dioica*, and in the Australian *Plagianthus* complex of the tribe Malveae.

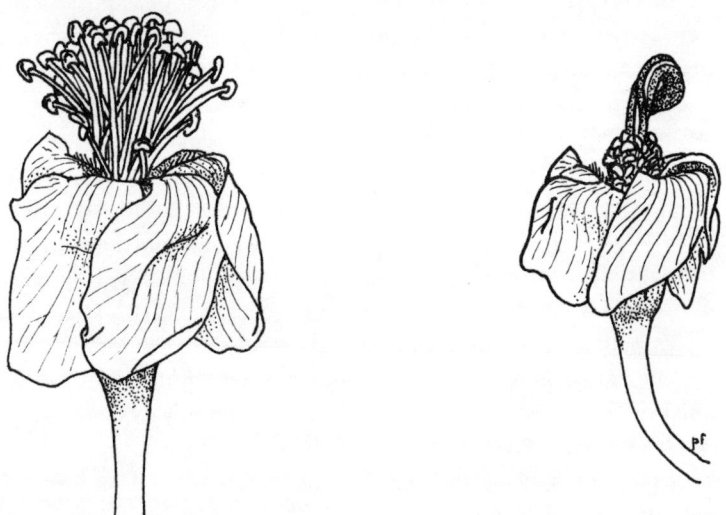

Fig. 49. Staminate (left) and pistillate (right) flowers of *Hampea nutricia*. (Reprinted from Fryxell, 1969*d*, fig. 2.)

In the case of *Hampea*, the dioecy is expressed morphologically by the total suppression of the pistil in the flowers of staminate plants and by the partial suppression (vestigial nature) of the anthers in flowers of pistillate plants. In the latter flowers, microsporogenesis breaks down at an early stage, and the anthers are indehiscent, pallid, and lacking mature pollen grains. Moreover, the filaments are much shorter in pistillate flowers than in staminate ones (Fig. 49).

An apparently similar condition exists in *Cienfuegosia rosei* (and, in the Malveae, in some species of *Callirhoë*, for example) that it is tempting to consider as transitional to the situation in *Hampea*. I believe, however, that this condition is a different pattern of floral dimorphism, that is, gynodioecy (or male sterility, or pollen abortion), instead of a transitional stage to the kind of dioecy expressed in *Hampea*. In gynodioecy as expressed in *C. rosei*, there is no reduction in the development of the pistil; all plants have normally developed (and functional) pistils. Pistillate plants simply have nonfunctional anthers resulting from a failure of microsporogenesis. That is, *if* the gynodioecy of *C. rosei* is a transitional step to dioecy, it is occurring by the loss of male fertility first, since there has yet been no loss of female fertility. In *Hampea*, on the other hand, dioecy was achieved by the loss of female fertility *first* (since the pistil has not a vestige remaining in staminate flowers) and the loss of male fertility subsequently (since a vestigial androecium remains in pistillate flowers). This interpretation is supported by the case of *Hampea tomentosa*, a dioecious species, in which pistillate flowers will rarely produce a few apparently normal pollen grains. If this condition represents a transitional state in *Hampea* (species of *Hampea* sect. *Standleya* having perfect flowers), then dioecy was evidently achieved by way of androdioecy, not gynodioecy.

The androecium of the Malvaceae is one of the distinguishing characters of the family (Fig. 50). The filaments of the numerous stamens are fused into a staminal column that surrounds the ovary and also surrounds most of the style or styles. The filaments have a free portion apically emerging from the column, each bearing its own anther. This general pattern is remarkably constant throughout the family. Except for the tribe Malveae, the apex of the staminal column terminates in five teeth, and the filaments all arise from below these teeth.

The Gossypieae follow this general pattern but show certain signif-

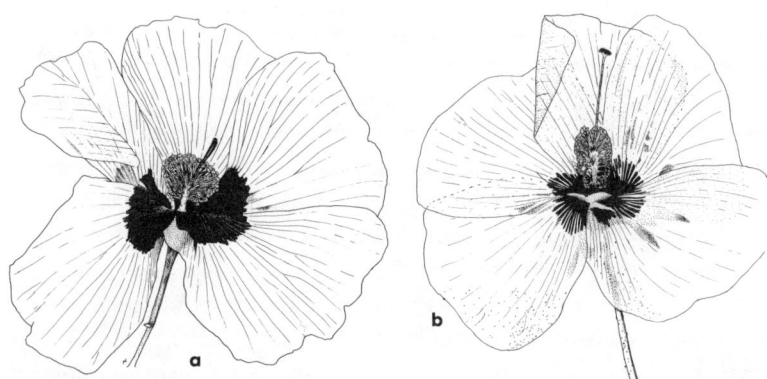

Fig. 50. Flowers of *Cienfuegosia hildebrandtii* (left) and *Cienfuegosia affinis* (right), illustrating decurrent and capitate stigmas. (Reprinted from Fryxell, 1969c, figs. 17a and 21b.)

icant variations. In certain species of *Cienfuegosia* (for example, *C. digitata* and *C. drummondii*), the emergent filaments and their anthers are more or less clearly grouped into five fascicles instead of being distributed uniformly around the circumference of the column. The staminal fascicles are opposite the petals and alternate with the sepals. There is sometimes a suggestion of a secondary division into ten instead of five fascicles.

Generally, however, the free stamens arise uniformly spaced or "randomly oriented" from the column, usually from throughout the length of the column or at least from the upper half of the column. (In the tribe Malveae, by contrast , the filaments arise only from or near the apex of the column.) The anthers are disposed in a three-dimensional space in a configuration that is often characteristic of a species. In some species (such as *Gossypium gossypioides* [Fig. 24] and *Cienfuegosia affinis*) the anther mass may be more or less cylindrical; more commonly it is ellipsoidal; in some (for example, *Cienfuegosia hildebrandtii* [Fig. 50A] and *Cephalohibiscus peekelii* [Fig. 6]) it is roughly spherical. These configurations are more evident in a fresh flower than in a pressed specimen, where they are only imperfectly represented.

Hampea is specialized in its floral morphology, and this specialization is evident in the androecium. In the perfect-flowered species such as *H. rovirosae* the staminal column is relatively short and the free filaments relatively long; the relative proportions of the flower parts are such that the stamens are mostly erect and the anthers are in close proximity to the stigma. In the staminate flowers of the dioecious species (for example, *H. nutricia* [Fig. 49]), the column is even shorter (almost absent), and the long filaments are free for almost their entire length. This form results in the "shaving brush" phenotype that is well known in several bombacaceous genera (such as *Pachira* and *Bombax*) and which in the past has resulted in the erroneous placement of *Hampea* in the Bombacaceae instead of the Malvaceae. If, however, one examines the pistillate flowers of the same species, their malvaceous nature becomes more apparent. In the flowers of pistillate species of *Hampea*, the pistil is fully developed and the androecium is vestigial. The staminal column and the free filaments are relatively short, so the anthers are grouped in a compact cluster around the base of the style and stigma, which emerges from and overtops the androecium. The anthers are indehiscent and sterile, having had their development arrested at an early stage. In either unisexual or bisexual flowers of *Hampea*, the five teeth at the apex of the staminal column, though not prominent, can be demonstrated, which again reinforces the placement of *Hampea* in the Malvaceae rather than the Bombacaceae.

The pigmentation of the staminal column shows some variation. It is pallid in most species, but in a few (for example, *Cienfuegosia somaliana*) it is dark-pigmented, evidently as an extension of the dark pigmentation found in the basal petal spot. Those species with pigmented androecial tissue generally have the petal spot larger or more prominently developed than average.

Pollen grains of the Gossypieae are typical of malvaceous pollen grains generally: they are relatively large, with diameters in the range of 50–150 microns, spheroidal, and spinose. The spines are numerous and commonly on the order of 8–10 microns in length, although the range includes spines 3–20 microns long, on the basis of the scattered reports that are in the literature. The number of apertures in the pollen grain is variable. Some species (*Lebronnecia kokioides* and *Gossypium stocksii*) have as few as three to five apertures per grain. Other species (*Gossypium barbadense* and *Hampea* spp.) have five to fifteen apertures, and still others (*Cienfuegosia affinis, C. drummondii*, and probably *Thespesia* spp.) have more than twenty apertures per grain. However, palynological observations on members of the Gossypieae are incomplete, scattered through the literature, and not always made in ways that may be compared with each other. A critical palynological study of the group has yet to be made. Pollen grains are so large and heavy (and often sticky) that they are not effectively, if at all, transported by the wind but require pollen vectors such as insects to carry pollen from one flower to another.

The ovary in the Gossypieae, as in all mallows, is superior, is situated within the staminal column, and has axile placentation. Unlike some mallows, however, the carpels are wholly fused throughout their length and are only three to five in number; there are rarely only two carpels per fruit in *Gossypium trilobum*. The style extends from the apex of the ovary up through the staminal column, from which it emerges at the top, more or less in proximity to the anthers. The style is single, the contributions from the several carpels being fused (connate) throughout their length, except that the styles are sometimes slightly divided apically in species of *Cienfuegosia* sect. *Friesia*.

Stigmas in the Gossypieae may be either capitate or decurrent (Fig. 50). Capitate stigmas are found only in species of *Cienfuegosia* in sections *Robusta, Paraguayana,* and *Friesia*. These species tend to have relatively long styles that place the stigmas too far from the androecium to permit ready self-pollination, presumably a character complex promoting outbreeding. However, in *C. argentina* the apically divided style permits the style branches to recurve and effect self-pollination, if cross-pollination has not happened. Those species having capitate stigmas generally have dark maroon pigmentation in the stigmas in contrast with the pallid style.

In the remainder of the species of *Cienfuegosia* and in all of the other genera of the tribe the stigmatic area is decurrent on the style, although in *Cephalohibiscus peekelii* the stigma is intermediate in form. There are as many stigmatic lobes as there are carpels, and these lobes may be slightly twisted. The entire style and stigma that emerges from the androecium generally has a clavate conformation. The length of this style varies; the stigmatic surfaces may be placed in among the anthers (promoting self-pollination) in species such as *Cienfuegosia heterophylla* and *Hampea rovirosae*, or the stigmatic surfaces may be held far above the anthers (promoting outbreeding) in species such as *Gossypium tomentosum* and *G. harknessii*.

The Fruit

FRUITS are capsular in the Gossypieae and, like the ovaries from which they develop, are three- to five-celled. In *Thespesia populnea* (Fig. 33), *T. grandiflora, T. patellifera,* and probably a few other species of

Thespesia the fruits are indehiscent. In other species of *Thespesia* and in all the other genera of the tribe, however, they are dehiscent capsules. They may be stout and woody, as in *Hampea tomentosa*, *Kokia* spp., and *Thespesia lampas*; they may be less substantially constructed, more coriaceous than woody, as they are in most species of *Gossypium*; or they may be relatively fragile, as they are in several species of *Cienfuegosia* (Fig. 7).

The fruits are sometimes greatly elongated (as in *Cephalohibiscus peekelii* [Fig. 6], *Cienfuegosia welshii*, and *Hampea platanifolia*), are generally ovoid (in most species) or obovoid (*Hampea* spp.), and are sometimes reduced to small, subspherical fruits (*Hampea sphaerocarpa* and *Cienfuegosia sulfurea*). In the indehiscent-fruited species of *Thespesia* the fruits are strongly oblate in form.

Typically there are many seeds per locule in the Gossypieae, but this number is reduced to one in some species, such as *Kokia* spp., *Lebronnecia kokioides*, and *Hampea sphaerocarpa*, or two in *Gossypium triphyllum* and *G. areysianum*, for example.

In most representatives of the tribe the mature fruit is glabrate, except in the genus *Hampea*, in which all species have pubescent fruits that are densely covered with "granular" or "mealy" stellate hairs. The color of these fruits in *Hampea* is characteristic of the different species. A similar but reduced type of indumentum occurs in *Lebronnecia kokioides* and *Thespesia lampas*. A few other species of the tribe also have pubescent fruits, with soft puberulence in *Gossypium australe* and *G. triphyllum*, with antrorsely oriented strigose hairs in such species as *Cienfuegosia affinis*, *C. gerrardii*, *C. welshii*, and *C. subprostrata*, and with dense stellate pubescence in *Thespesia garckeana* (Fig. 42). Upon dehiscence of the capsules, the suture of dehiscence often shows a fringe of distinctive hairs along the inner margin of dehiscence, usually in *Hampea*, *Gossypium*, some species of *Thespesia* (such as *T. lampas*), and sometimes *Cienfuegosia* (in section *Cienfuegosia*, for example, but not in section *Robusta*). Those hairs are weakly expressed in *Kokia* and *Lebronnecia*.

In some species (for example, *Gossypium somalense* and *G. gossypioides*) abscission takes place upon maturity of the fruit at the base of the pedicel, and the entire fruit, bracts included, falls. More commonly the fruits are persistent for a long period after maturity, and only individual seeds are dispersed. In such cases the mature carpel walls may flare widely (as in *Hampea nutricia*, *Gossypium harknessii*, *Cien-*

fuegosia hearnii, and *Cephalohibiscus peekelii*), or they may flare very little (as in *Hampea trilobata*, *Gossypium thurberi*, *Thespesia lampas*, and *Kokia* spp.), resulting in relatively prompt or relatively delayed release of the seeds, respectively. Intermediate situations, of course, exist.

Seeds and Seed Hairs

THE seeds of the Gossypieae are of special interest because in the cultivated species of *Gossypium* they have been the source of the culturally and commercially important cotton fibers from prehistoric times to the present. So first let us consider the seed hairs.

With the exception of the genus *Hampea* and a few isolated species in the other genera, species of the tribe Gossypieae characteristically have hairy seeds. The seed hairs are outgrowths of the epidermis of the seed coat. Each hair is a single cell, the base of which is embedded in the epidermis; it is many times longer than wide—up to about fifteen hundred times longer than its diameter in cultivated cottons. These seed hairs vary in color, principally ranging from a rich chestnut brown through lighter shades to the pure white that has been selected for in commercially cultivated cottons. To some extent, various shades in this range of color are characteristic of individual species. For example, seeds of *Gossypium lobatum* and *G. harknessii* have silvery gray hairs, whereas *G. aridum* and *G. armourianum* have brown hairs. Mutant forms of *Gossypium hirsutum* exist that have greenish fibers. The fibers quickly fade to a nondescript gray or tan when the fruit matures, however, because the green pigment is unstable in sunlight; this color can only be preserved by storage in the dark.

In addition to all of the species of *Hampea*, a few other species have the seed hairs more or less suppressed. Seeds are wholly glabrous in such species as *Thespesia lampas*, *T. grandiflora*, *T. beatensis*, and *T. cubensis*. The seed hairs are very short and often sparse, resulting in a seed that is puberulent or subglabrous, in such species as *Thespesia thespesioides*, *Cienfuegosia affinis*, *Gossypioides kirkii*, *Gossypium thurberi*, *G. trilobum*, *G. klotzschianum*, *G. davidsonii*, *G. populifolium*, and *G. cunninghamii*.

In the majority of species, however, the seeds are densely and copiously lanate, so much so that the surface of the seed coat is quite invisible. These hairs may be disposed in several different ways. In species such as *Gossypium harknessii*, *G. armourianum*, and *Cienfuegosia drummondii* the seed hairs are appressed to the seed coat so tightly that the seeds at first glance do not appear to be hairy; they may be demonstrated to be hairy only by teasing the hairs loose from the seed surface with a needle.

In other species, such as *Thespesia danis*, *T. populnea*, *T. garckeana*, *Gossypium bickii*, *G. lobatum*, *Cienfuegosia heterophylla*, and *C. argentina*, the seed hairs are also appressed to the surface of the seed but not so tightly that the hairy nature of the seed is not evident. The hairs are often wavy or crimped and curled against the surface of the seed in a coordinated fashion.

Species such as *Gossypium triphyllum*, *G. raimondii*, *Cienfuegosia rosei*, and the cultivated cottons have seed hairs that because of their length, curliness, and density are relatively undisciplined. Their seeds are well described as "cottony."

In *Lebronnecia kokioides* and species of *Cienfuegosia* sect. *Garckea* (Fig. 7) the seed hairs are of comparable length and density to the "cottony" group just described, but the hairs are relatively straight and emerge from the seed in a more or less tangential orientation. The seed and its hairs thus have a "pinwheel" appearance. The extreme case is found in *Gossypium australe*, *G. nelsonii*, and *Cephalohibiscus peekelii* (Fig. 6). In these three species the seed hairs are perfectly straight and fully patent, sticking out in all directions from the seed.

Hampea has already been noted as exceptional in having glabrous seeds in all species. They are not only glabrous but also arillate. The aril in *Hampea* is soft and fleshy, snow-white when fresh, and tasteless. It partially encloses a portion of the seed and nearly equals it in size. In those species of *Hampea* in which the fruit does not flare widely upon maturation (such as *H. tomentosa* and *H. platanifolia*), the aril is somewhat smaller. Elsewhere in the Gossypieae, arils are found only in *Gossypium cunninghamii* and *G. populifolium*, although they are not well developed. These species also have subglabrous seeds.

The seed coat is a relatively indurate structure, impregnated with lignins and tannins, in most species of the tribe. In freshly matured seeds the seed coat is impermeable to moisture, and as a result the seed has a degree of dormancy (sometimes called "hard-seededness"). The dormancy is "broken" and the seed becomes germinable after weathering or after the seed coat is physically damaged. There are exceptions to this pattern, however, although the matter has not been well studied except in the cultivated cottons. Relatively fragile seed coats are found in *Gossypium australe* and *G. lobatum*, but I am not aware that careful studies have been made of their germinability, dormancy, or longevity. There has been strong selection practiced in the agricultural cottons against impermeable seed coats, with the result that present-day cultigens have lost the hard-seededness that characterized their wild forebears. The farmer can plant his seeds and expect germination to follow promptly, not months or years later when bacterial action has finally broken down the impermeable seed coat so that moisture can reach the embryo.

The matter of seed longevity was mentioned in the preceding paragraph. As one might suppose, seeds that have impermeable seed coats and hence seed-coat dormancy have relatively long-lived seeds. Only seeds of *Gossypium* spp. have received much experimental study in this respect, but they are presumably typical of much of the tribe. Seeds of *G. thurberi* lying in the soil retain dormancy and viability for at least twelve years (Endrizzi, 1974). With favorable storage conditions (cool and dry) some seeds of *Gossypium* spp. retain their viability over periods of ten to twenty (even thirty) years. On the other hand, seeds of *Hampea* spp., based on very limited data, seem to lose their viability within approximately a year of maturation, and in the case of *Thespesia grandiflora* seed longevity is evidently to be measured in months. Fuller studies of this subject will be of considerable interest, especially if they are related to the habitats in which the plants grow and the reproductive cycles that occur in nature.

The anatomy of the seed coat in the Gossypieae has been studied in several species of *Gossypium* and a few other species of the tribe and contrasted with other representatives of the Malvaceae (Reeves, 1936; Corner, 1976). The seed coats are characterized by a thick palisade layer of long, columnar cells that, although it is only one cell layer thick, occupies more than half the thickness of the seed coat. This palisade layer is found in most genera of the Malvaceae but is seemingly better developed in the Gossypieae than elsewhere in the family. Immediately external to the palisade layer is a heavily lignified, unpigmented layer called the crystal layer. The presence of a crystal layer appears to be unique to the Gossypieae. A tissue overlying the crystal layer and underlying the epidermis, called the outer pigment layer, is pigmented and suberized and includes integumentary vascular bundles. The presence of these integumentary bundles appears, again, to be unique to the Gossypieae. Presumably the suberization of the outer pigment layer and the lignification of the crystal layer are related to the impermeability of the seed coat and resultant hard-seededness, but I am not aware that this relationship has been explored experimentally. It should be emphasized that these anatomical studies of the seed coat have principally concerned species of *Gossypium* (and also *Thespesia populnea*), but that other representatives of the tribe have not yet been studied.

Seeds of the Gossypieae vary rather widely in size, from the small seeds of *Cienfuegosia yucatanensis* and *Gossypium sturtianum* (3–4 mm long) to the large seeds of *Lebronnecia kokioides* and *Thespesia grandiflora* (10–12 mm long). The embryo that is contained within the seed is the distinctive structure that was described at the beginning of this chapter.

4

The Origin and Spread
of the Tribe

*In the "evolution game" which it is playing, a species has to
contend with unforeseen eventualities which the future may
bring—a new parasite, a new predator, possibly an Ice
Age.... The necessity for survival of the species is that [it
contain] individuals that succeed in leaving offspring in the
prevailing environmental conditions. There are quite a wide
variety of "strategies" by which a species can attempt to meet
this requirement; in fact, they affect different levels of [biologi-
cal] organization, so the species is playing several different
games simultaneously.*

C. H. Waddington, 1965

EVEN a cursory glance at the geographical distribution of the plants of
the Gossypieae (Fig. 51) reveals two notable characteristics of their
distribution: (1) they are plants of the tropics and subtropics and, except as cultigens, are essentially excluded from temperate climates,
and (2) they tend to be plants of the Southern Hemisphere.

The first observation scarcely needs further comment, except to
note that it has implications concerning the phytogeographic history of
the group. The implications of the second observation, however, require a more detailed exploration.

The gross elements of the geographical distribution of the tribe are
as follows:

1. South America—a major development of *Cienfuegosia* and a minor
   development of *Gossypium*
2. Middle America—an extension of the South American development
   of *Cienfuegosia*, major developments of *Gossypium* and *Thespesia*,
   and the principal development of *Hampea*
3. Africa (including Arabia)—major developments of *Cienfuegosia*,
   Gossypium, and *Thespesia*, and the entire development of the small
   genus *Gossypioides*

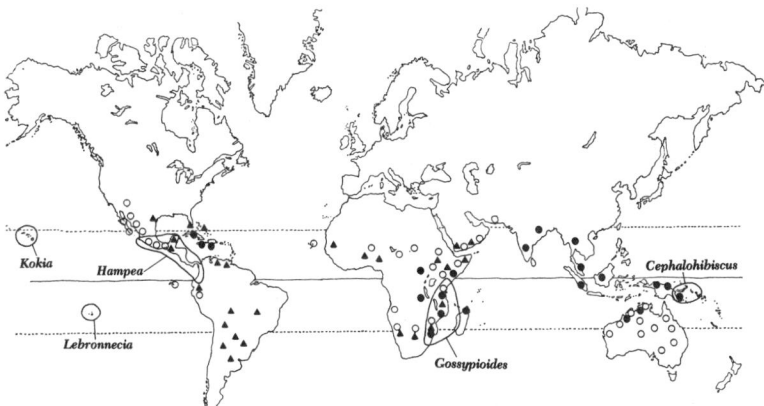

Fig. 51. Geographical distribution of the genera of the Gossypieae. ○ = *Gossypium*; ▲ = *Cienfuegosia*; ● = *Thespesia*. Excludes the tetraploid species of *Gossypium* (see Fig. 53), the cultivated diploid species of *Gossypium*, and the littoral species of *Thespesia*, *T. populnea*, and *T. populneoides*.

4. Australasia—a major development of *Gossypium* in Australia and of *Thespesia* in New Guinea

5. Oceania—a minor development of *Gossypium*, a major development of *Thespesia* (in New Guinea), and scattered development of the small genera *Cephalohibiscus*, *Lebronnecia*, and *Kokia*.

The detailed knowledge that is available of the genus *Gossypium* has given rise to a general consensus that it is an ancient genus. The broader scope provided by a consideration of the entire tribe Gossypieae substantiates this view. The taxa included in the tribe are highly diversified and are separated by important geographical and (where known) reproductive barriers that imply a long evolutionary history. The principal exception to this statement is the genus *Hampea*, the species of which are less highly differentiated among themselves and less widely dispersed geographically, from which we may conclude that *Hampea* is perhaps a younger genus than the other large genera of the tribe.

It is difficult to discuss the actual age of ancient taxa such as this in the absence of any fossil data, and it is not considered particularly meaningful to do so, except in the relatively broad context of what is known about palaeoclimates. We may, however, collate (1) the known restriction of these plants to essentially frost-free areas, (2) the known climates of earlier geologic ages that were mild to much higher latitudes than they are today, (3) the inferred austral development of this group of plants, and (4) the time of breakup of the ancient Gondwana landmass and from this collation draw the not improbable conclusion that this tribe, as a distinct lineage, is perhaps of Cretaceous age.

On what basis do we infer an austral origin for the group? The evidence is hardly conclusive, but the following data are considered significant: There is a clear geographical tie (of remote origin) between the two subgenera of *Cienfuegosia*, the one South American, the other African. There is a geographical tie (somewhat less clear) between the subgeneric taxa of *Gossypium* in Australia and in Africa. There is the suggestion of a remote taxonomic relationship between *Cienfuegosia* sect. *Articulata* from South Africa and *Thespesia* sect. *Lampas* from Australasia. Finally, the several taxa of Oceania (including the New Guinea *Thespesias*, *Cephalohibiscus*, *Lebronnecia*, *Kokia*, and perhaps extending to the Central American *Hampea*) have ties to one another that point south.

How the last group may have been dispersed to its present distribution is a matter for speculation. It is interesting to note, however, that the four species of *Thespesia* endemic to New Guinea are essentially confined to that single landmass, whereas the monotypic *Cephalohibiscus* occurs not only on New Guinea, but also on New Ireland, Bougainville, and Santa Isabel across oceanic barriers of several hundred kilometers. It would seem that there is a greater potential for "island hopping" in the one case than in the other, although the possibility of dispersal by man in the case of *Cephalohibiscus* cannot be entirely discounted (cf. Ulbrich, 1935). In any case, such differences need to be taken into account in attempting to explain distributional patterns. However, the taxa concerned are so rare (*Lebronnecia* and *Kokia* being on the verge of extinction) and so widely scattered that speculation on the history of their dispersal cannot be expected to be very fruitful.

The first observation to be made about the distribution of the Gossypieae—indeed, it is an obvious fact—is that the tribe is tropical and subtropical in distribution. Other observations are nearly as obvious: that the tribe is worldwide in distribution and that in certain particulars its distributional pattern shows a relationship to marine dispersal. But further generalizations are less obvious and demand closer scrutiny and fuller exposition. The entire question of the distribution of the Gossypieae raises questions about the means of dispersal of the members of the tribe that must first be considered.

As van der Pijl (1969, p. 1) has pointed out, it is essential to distinguish between "actual dispersal as studied in the field, and the structural basis needed to attain this dispersal." In the Gossypieae, data of the former kind are unfortunately rare, and much of the discussion will of necessity have to be based on structural considerations, supplemented where possible with information on actual dispersal mechanisms.

Seed Hairs and Seed Dispersal

DISPERSAL in the Gossypieae is virtually synonymous with seed dispersal, although dispersal of whole fruits occurs in some species of *Thespesia*. None of the members of the tribe has developed vegetative propagules.

One of the notable features of these seeds is the dense coat of hairs growing out of the seed coat, which is well known as the commercial crop of cotton produced by certain species of *Gossypium*. This dense coma occurs in all of the genera of the tribe except *Hampea*, which has totally glabrous seeds. Several individual species among the other genera, however, also have seeds that are either totally glabrous (for example, *Thespesia lampas, T. grandiflora*, and *T. cubensis*) or subglabrous (for example, *Gossypium thurberi* and *Cienfuegosia affinis*). The latter type does not have a coma developed sufficiently for significance in dispersal.

A few species have patent hairs (for example, *Gossypium australe, Cienfuegosia hildebrandtii*, and *Cephalohibiscus peekelii*) that stick straight out from the seeds. Most of the hairy-seeded species, however, have seed hairs that are wavy or crisped and loosely appressed to the seeds. A few (such as *Cienfuegosia drummondii, Thespesia danis*, and *Gossypium harknessii*) have hairs that are so tightly appressed to the seeds that they appear hairless, although closer examination shows the hairs to be dense when teased loose from their tightly appressed condition.

It is difficult to be dogmatic concerning the general significance of these seed hairs in seed dispersal. The character is certainly firmly established in the Gossypieae and occurs in several genera of the Hibisceae, but it is unusual elsewhere in the Malvaceae. A character as firmly established as the seed hairs of the Gossypieae ("evolutionary canalization," in Stebbins' [1974] terminology) is presumed to be a long-established character and one that has an adaptational basis. It is not necessary to specify in detail what that adaptational basis is (or may have been) to recognize that its basis is indeed adaptational. Cronquist's (1968) oft-repeated phrase, "If [the character under consideration] has any functional significance, it remains to be discovered," contributes little to our understanding of botanical problems. To accept such cases as adaptational may be on faith rather than by a clear demonstration of fact, but at least that faith is *rational* faith based upon the firmly established Darwinian view of biology, not a know-nothing faith requiring a fresh proof for each case. The point overlooked by Cronquist and expounded by Stebbins is that the adaptational basis for a given trait need no longer be extant for the trait to have survived in extant genotypes; it may have become genetically fixed at an earlier time under different evolutionary circumstances.

As a manufactured, facetious example, the early ancestors of the Gossypieae may have had the genetic basis for hairy seeds fixed (through selection) in their genotype because of the dispersal advantage conferred from the fact that some extinct reptile used these seeds for nest building. That the reptile (perhaps a pterodactyl?) is long since extinct does not alter the fact that the genetic basis for comose seeds was fixed in the germplasm of this line at an early date and is today present for initially sound adaptational reasons, coupled with a resultant narrowed genetic base, even though the mists of time may hide the specific details from our understanding. Of course the functional significance "remains to be discovered." It may well no longer exist or even be discoverable. The argument from ignorance is inadequate to dethrone selection and to replace it with mystical forces such as mutation pressure.

In the case of the Gossypieae, the very firmness of the establishment of the hairy-seeded condition suggests its long-term establishment and further suggests that its original adaptational "reason" may well have been lost from sight because of its great age. Indeed, clear stories of adaptation are more profitably sought among the glabrous-seeded exceptions, on the assumption that that condition has arisen more recently as a specialization.

The Aril as a Decoy

THE most obvious exception to the hairy-seeded condition is the genus *Hampea*, in which all of the species have glabrous, arillate seeds. There can be little doubt that the glabrousness is closely related to the arillate condition and that both are involved in the matter of seed dispersal. Seed dispersal in *Hampea* is exceptional and relatively specialized for the Gossypieae. The arils are snow-white and contrast notably with the dark, shiny seeds. Their showy presentation is evidently a means of attracting potential disseminators, presumably birds. This is a relationship typical of arillate plants generally. Seeds of *Hampea* that fail to attract disseminators and instead fall to the ground germinate promptly (without dormancy) in large numbers immediately below the parent tree. Few of these can survive; wastage of seeds and seedlings is high. Effective dispersal can occur only by some other means.

The aril in *Hampea* is a soft, fleshy structure without any protective covering. Consequently, it dries out, shrivels up, and loses its eye-catching white color rather promptly after maturation of the fruit. It also falls out of the capsule rather promptly after being presented to potential disseminators for only a limited time (a few days at most) in the freshly opened capsule while the aril is fresh and succulent. While thus presented, the seeds are still up in the crowns of the trees and would be available principally to birds, although certain arboreal mammals could conceivably function as dispersal agents. Anecdotal information is available, however, and indicates that the seeds of *Hampea* (specifically *H. nutricia* and *H. appendiculata*) are attractive to birds, and it is logical to look to birds as disseminators of *Hampea*. For example, Dr. Gordon Frankie (personal communication) has observed two species of toucan in Costa Rica feeding heavily on freshly ripened seeds of *Hampea appendiculata*. A field study of the joint distribution of *Hampea* spp. and of the birds that feed upon (and disseminate) their arillate seeds would probably be enlightening.

Corner, in a series of papers (1949, 1953, 1954), has developed an elaborate theory on "the origin of the modern tree" that is based on observations of arillate seeds of tropical plants and that he has chosen to call the "durian theory." The name derives from the striking and widely renowned fruits and seeds of *Durio zibethinus* Murray of the Bombacaceae, although Corner draws many of his examples from the Leguminosae and other families. Corner's theory has been challenged by Parkin (1953), van der Pijl (1952, 1969), and others and has never been widely accepted (cf. Eyde, 1976), although he continues to expound it (for example, Corner, 1976). Its elaborate reasoning rests firmly on the premise that the presence of an aril or arilloid is a character that is primitive for the angiosperms. Specifically, Corner (1976, vol. 1, p. 180) derives the Malvaceae from "ancestral Bombacaceae" and states that the two families "practically merge, as the durianologist would expect," in reference to the arillate seeds of *Hampea*. Such an interpretation does not take into account the crucial evidence of chromosome numbers (which involve relatively low base numbers in the Malvaceae and high polyploids in the Bombacaceae) and serves only to "support" the curious views of the durianologist. Strangely, the principal critics of the theory, with the exception of Endress (1973), have had no quarrel with this cornerstone assumption, although botanists generally are divided on the question of whether the arillate condition is primitive or derived. Sporne (1954) advances a statistical argument in support of the view that it is primitive; Cronquist (1968, p. 114), on the other hand, says that the aril and sarcotesta "are obviously adaptations relating to seed dispersal" and therefore derived; and Endress (1973) presents detailed morphological evidence that the aril is derived.

I wish to support and elaborate Cronquist's and Endress' view, because the facts that have been presented by Corner are very striking and draw attention to a significant phenomenon in tropical botany. I believe, however, that the facts may be interpreted differently than Corner and van der Pijl, for example, have done, and that the alternative interpretation is more persuasive.

The facts are that arils (the term is here used in a broad sense, because my interest is in the adaptational phenomena, not in morphology) are widespread among tropical plants of many families; that the aril is brightly colored, usually red but also white or yellow, in contrast to a dark seed; that the aril is fleshy in contrast to a hard seed; that the arillate seed is often presented at maturity on an elongated (pendent) funiculus; and that the fruit containing these seeds is often indurate, spiny, or otherwise repellent, being foul smelling in the durian. That this elaborate complex of characters should be repeated with such fidelity among so many relatively unrelated angiosperm families (listed by Corner) is indeed a striking phenomenon. Corner is to be thanked for drawing our attention to it so forcefully.

Two alternative views may be employed to account for a phenomenon of this sort. On the one hand, one may hold that such a widespread distribution of a character results from the fact that the character is primitive and that it characterized the early progenitors of the angiosperms. It has been lost in many groups but hangs on as a relict, indicative of the earlier history of the several lineages in which it is found, especially in tropical areas that have been subjected to the least climatic change. This is Corner's view. Those plants that have no arils or in which the arils are not fully developed are regarded by Corner as products of degeneration or reduction series from arillate types. Even if true, this view fails to answer the basic question of how and why this remarkable character complex arose in the angiosperms.

Alternatively, one may see this character association as one too complex to have remained intact over the period of time through which the families and orders of angiosperms have evolved. Rather, such a character complex could have been achieved and maintained only from the overriding imperative of adaptation. Parallel evolution is the watchword. That the parallelism should be as striking as it is in this case indicates the selection force to be one that is strong and that cuts across taxonomic groups.

I find that the former view, as applied to the arillate seed complex described by Corner, strains my credulity. At the same time, the latter view can be found persuasive only if a plausible and self-consistent adaptational pattern can be proposed that accounts for the observations. I believe such a proposal is possible.

The character complex is, as Cronquist states, obviously an adaptational pattern relating to seed dispersal. Moreover, it relates to active agents of seed dispersal: birds, mammals, and, as van der Pijl emphasizes, ground-dwelling reptiles in the earlier stages of angiosperm evolution. The bright colors of the aril are clearly eye-catching to vision-oriented animals, especially where these colors contrast with the black seed coat. The prominent means of presentation, often on a pendent funiculus, emphasize this feature. The arils are succulent, presumably nutritious, and therefore attractive to chronically hungry animals.

The remarkable fact that Corner emphasizes, however, is that the attractiveness of the arillate seed is in sharp contrast to the unattractiveness of the fruit that produces it. The fruits of *Durio*, for example, are hard, spiny, and notoriously fetid. The fruits of most other arillate species have at least one such protective or repellent characteristic.

The adaptational pattern, therefore, seems to be as follows. The positively unattractive fruits serve to protect the *maturation* of the seeds, which are often relatively large and thus vulnerable to predation when immature. At maturity a turnabout occurs, and the seeds (and their arils) are presented in a showy and attractive manner, using one or more of the devices mentioned: bright colors, pendent presentation, and edibility of the aril (reinforced by an exquisite flavor in the durian). The aril thus serves as a "decoy" that is edible (and thus rewarding to the disseminator) but which is dispensable to the plant. The seed "goes along for the ride" and is thereby dispersed. This is evidently a successful evolutionary ploy, judging by its widespread occurrence in tropical plant families.

The arillate seeds of *Hampea* of the Gossypieae conform to this adaptational pattern. The fruits are not spiny or fetid, but they are hard, sometimes woody, and unattractive to animals when immature and protecting the developing seeds. The seeds lack the elongate funiculus (in common with other arillate members of the Malvales) but are nevertheless presented in a showy manner by the widely flaring capsules. In one species, *H. breedlovei*, the whole capsule is pendent on an elongated pedicel. The arils are snow-white when fresh and contrast notably with the shiny black seeds; the newly opened capsules are indeed eye-catching. The arils are soft and fleshy and quite edible, although they are rather tasteless to my palate. The arillate *Hampeas* thus extend Corner's list of arillate families by adding the Malvaceae.

The arillate condition, it seems to me, is best interpreted as an adaptational pattern of considerable complexity that is derived from a less specialized background. It is so interpreted in chapter 9. It seems evident that the aril is produced at some metabolic cost to the plant and that it must, therefore, have a comparable adaptational value to the plant to be maintained.

It is of interest to note the morphological similarity of *Hampea* and *Lebronnecia*. Yet arillate seeds occur only in the continental *Hampea* and not in the insular *Lebronnecia*, which suggests that the arillate pattern evolved relatively recently. This view, moreover, is consistent with the interpretation presented here that the protective fruit and the attractive aril are adaptations of the plant to guard it against seed predation by large animals. Such predators would not be expected to be present to exert selective pressure in the remote insular habitat where *Lebronnecia* occurs, in contrast to the abundance of such predators in the continental habitat of *Hampea*.

The adaptational pattern described for *Hampea* clearly indicates the adaptational advantage of seed glabrousness in these plants. The seeds are dependent for their dispersal on visual stimuli to keen-eyed birds. The specific stimuli involve the striking contrast between the dark seed and the snow-white aril. Pubescence on the seed coat would detract from this contrast. Indeed, the seeds not only are glabrous but also have developed a high gloss that accentuates the contrast.

The "Saltshaker" Capsule

So we find *Hampea* to be a rather special case. What about other species of Gossypieae in which the seed hairs have been eliminated or reduced in size or amount to an insignificant level? Examples include *Cienfuegosia affinis, Gossypium davidsonii, G. klotzschianum, G. thurberi, G. populifolium G. costulatum, G. trilobum, Thespesia thespesioides*, and *Gossypioides kirkii*. The first striking fact to note is that the group cuts across genera. This fact suggests the independent origins of the reduced seed hairs in the several genera, in response, perhaps, to some common selective pressure. What might that pressure have been? Another way to put the same question is to ask, What ecological traits have these plants in common?

There no doubt are many such traits to be named, but ones that stand out are (1) a shrubby growth habit, (2) a distribution in open habitats (desert or savanna), and (3) the possession of a "saltshaker" capsule, that is, one that is often rather woody and that opens only partially, thus releasing the seeds gradually over a period of many months in response to shaking motions. To this constellation of characters must be added a character common to most members of the Gossypieae (except *Hampea* spp. and a few others)—an impermeable seed coat, which confers a temporary dormancy. This seed coat dormancy is ultimately broken by weathering or abrasion of the seed coat. It may thus last a few months or several years, depending upon the history of the individual seed after its dissemination. In any event, the seed coat dormancy permits dispersal over a greater distance by delaying germination over a greater period of time. The saltshaker capsule character, of which subglabrous seed is an essential factor, spreads the dispersal itself over a period of time. More effective dispersal is thereby insured whether seed is released by wind action on the parent shrub, by the tumbling of broken branches (with capsules) downstream after desert deluges, or by the transport of capsules or branches by animal disseminators. Observational data are badly needed on the actual means of dispersal (especially to distant sites) of this type of plant. Such observations are not easily obtained, but it is nevertheless evident that the several characters that together make up the saltshaker capsule are an adaptive complex concerned with seed dispersal. The reduction of seed hairs in the species mentioned is an integral part of this adaptive complex.

There remain to be considered several other species with glabrous seeds, all in the genus *Thespesia**: *T. grandiflora*, *T. beatensis*, *T. cubensis*, and *T. lampas*. The last-named species is a close ally of the previously mentioned *T. thespesioides* but differs strikingly in its seed vestiture. *T. thespesioides* has densely short-pubescent seeds; *T. lampas* has shiny glabrous seeds. Both have the saltshaker type of capsules, however, and may simply have responded in a different though equally effective way to their dispersal requirements. *T. lampas* enjoys a status as an ornamental cultigen, which may or may not relate to the problem at hand.

Indehiscent Fruits Go 'Round the World

THE remaining glabrous-seeded *Thespesias* are quite different. They are more arborescent, and they tend to come from more mesic habitats. The most striking difference, however, is in the nature of the fruits. The fruits are more or less indehiscent and are filled with a spongy pulp that surrounds the seeds. Indehiscent fruits are also found in *T. populnea*, and almost indehiscent fruits in *T. populneoides*, species that have short pubescence on their seeds. It is clear in these cases that the dispersal element is the fruit rather than the seed.

The indehiscent, buoyant fruits of *T. populnea* are remarkable examples of dispersal by sea, as will be discussed later. The more immediate question is: Why are not the other species of *Thespesia* that have comparable fruit structures also dispersed by sea? The three Caribbean species, *T. grandiflora*, *T. beatensis*, and *T. cubensis*, are each confined to individual islands (although *T. grandiflora* has gone elsewhere in cultivation), and of the four New Guinea species, *T. multibracteata* is endemic to Ferguson Island, *T. fissicalyx* and *T. robusta* occur only on New Guinea proper, and *T. patellifera* crosses only the trivial gap from Papua to the islands of Goodenough and Normanby, distances across water of approximately thirty-three kilometers. The answer is not completely known, but it is suspected

---

*The seed vestiture of several of the New Guinea species of *Thespesia* is unknown. Since their fruit characters are in many respects similar to those described above, it is probable that their seed characters are also.

that the known ability of seeds of *T. populnea* to retain their viability after prolonged exposure to seawater is not shared generally by these other species. It is known, for example, that the seeds of *T. grandiflora* are relatively short-lived and retain their viability for little more than a month even under optimum storage conditions. They succumb quickly upon exposure to seawater, since they lack an impermeable seed coat.

In any case, it is clear that these species are *not* dispersed by sea but that their dispersal organ is the whole fruit. Given this adaptive syndrome, selective pressures will not be operative on seed-coat vestiture as a dispersal character, and other considerations will come to bear. Evidently these other considerations, which may well relate to the reproductive energy balance of these species, have tipped the balance in favor of glabrous seeds in most of the fruit-dispersed species.

"Tumbleseeds"

THE species with patent seed hairs include *Cephalohibiscus peekelii*, *Gossypium australe*, *Gossypium nelsonii*, *Cienfuegosia hildebrandtii*, and *C. heteroclada*. The first-named species is a rain-forest tree from several Melanesian islands that, as a monotypic genus, stands distinctly apart. Nothing is known of the means of seed dispersal in this species or of the relation of the distinctive seed hairs to it. The other three species are subshrubs of open (desert or savanna) habitats. It has been suggested that their distinctive seed vesture serves to make each seed a miniature tumbleweed or "tumbleseed," thereby very effectively dispersing seeds from the parent plant. Upon maturation, the capsules flare widely, releasing the seeds to wind or gravity and enabling them to tumble quickly away with the first breeze. Seed dormancy is evidently lacking, so germination can occur promptly whenever an adequate moisture supply is encountered. This seed dispersal mechanism has evidently been successful in the case of *G. australe* and *C. hildebrandtii*, both of which are more widely distributed than their closest allies.

It is rather difficult to suggest an adaptational basis for the tightly appressed seed hairs encountered in such species as *Cienfuegosia drummondii*, *Gossypium harknessii*, *G. turneri*, and *G. armourianum*. It seems, however, that these species found comose seed disadvantageous (presumably in relation to dispersal, but not necessarily so). Instead, they hit upon the evolutionary strategy of appressing the seed hairs tightly against the seed to produce a pseudohairless condition rather than dismantling the genetic apparatus for producing seed hairs and *actually* denuding the seeds. This is a good example of Stebbins' "adaptive modification along the lines of least resistance." What these selection pressures may have been remains open to conjecture. These species have not developed the saltshaker capsule syndrome but have widely flaring carpels that drop their seeds to the ground very promptly upon maturation.

In the case of *C. drummondii* the dispersal pattern evidently involves bird disseminators, specifically representatives of the Charadriiformes. *C. drummondii* is a low perennial herb that commonly grows in wet meadows, often in heavy, saline soils. The seeds are nutritious and attractive to seed-eating birds. Presumably their pseudohairless condition makes them more attractive as food than does a dense coma. Their impermeable seed coat permits them to resist the digestive juices of the birds' digestive tract, should the seeds survive the action of the birds' gizzard. Their size (5–6 mm) is exactly in the range indicated by Proctor (1968) from experimental data that permits some of them to be trapped in a corner of the gizzard for a protracted period of time (up to several weeks). Those that are so trapped escape the grinding action of the gizzard and are eventually regurgitated intact— dispersed and ready for germination. The facts that both the plants and the birds frequent similar habitats, that the birds are long-distance migrants between the Northern and Southern hemispheres, and that *C. drummondii* has a disjunct distribution between the Northern and Southern hemispheres at comparable latitudes (about 28° N and S, in southern Texas and in northern Argentina and Paraguay, respectively) tend to support the interpretation of *C. drummondii* as a bird-dispersed plant. The appressed seed hairs are viewed as a specific adaptation to this mode of dispersal, rendering the seeds more attractive to the disseminators.

Saltwater Dispersal and the Littoral Habitat

Now that the exceptions have been discussed, we may turn our attention to the basal phenotype that is typical of the tribe Gossypieae: comose seeds. This group includes *Lebronnecia kokioides*, *Kokia* spp., *Gossypioides brevilanatum*, *Thespesia garckeana*, *T. gummiflua*, and most of the species of *Gossypium* and *Cienfuegosia*.

As stated at the outset, the general significance of these distinctive seed hairs for seed dispersal is difficult to state with finality. They may enable the seeds to adhere to the fur of passing animals, or they may serve to trap and hold moisture for germination in an arid climate. It has even been suggested that the fibers are attractive to birds as nesting material, which has been demonstrated at some localities. Other possibilities can be suggested, but in the absence of critical field observations the preparation of such a list is a relatively fruitless exercise.

One aspect of this question, however, deserves a closer examination. It was stated earlier that the distribution of the tribe as a whole showed a pattern suggestive of marine dispersal. Most of these cases (except for the remarkable distributional pattern of *Thespesia populnea*) involve species with comose seeds. We find, for example:

1. *Cienfuegosia yucatanensis* distributed around the shores of the Gulf of Mexico in a few coastal sites (Fig. 52): Yucatán, the Florida Keys, the Cuban Cayos, and at least two of the Bahama Islands

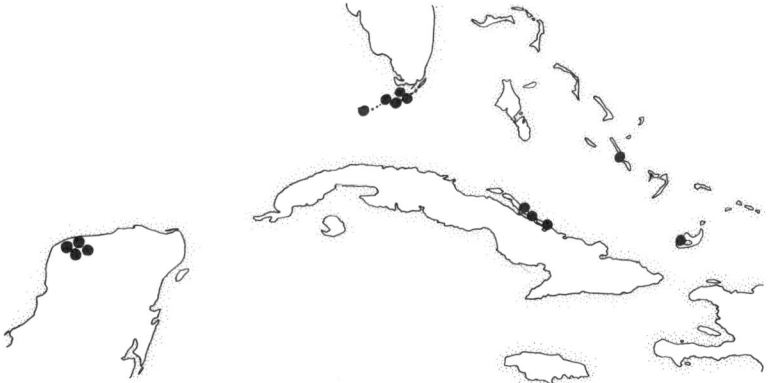

Fig. 52. Geographical distribution of *Cienfuegosia yucatanensis*.

2. *Cienfuegosia heterophylla* principally distributed in Venezuela and Colombia but extending to offshore islands like Aruba, Margarita, and as far as St. Thomas in the Virgin Islands
3. *Cienfuegosia digitata* being separated by the Atlantic Ocean (at that ocean's narrowest point) from its nearest relative, *C. heterophylla*, and from all the rest of the other sixteen species in its subgenus
4. *Gossypium darwinii*, of the Galápagos Islands, evidently dispersed to and among these islands by sea currents from the South American mainland, where its close relative *G. barbadense* is fully developed
5. *Gossypium hirsutum*, the most remarkable example of all, distributed throughout the Pacific Ocean (Socorro Island, the Marquesas,

Samoa, Tahiti, Wake Island, and northern Australia) wherever ocean currents carried it from its Central American origin, and also found in littoral habitats around the Caribbean Sea and the Gulf of Mexico

6. *Gossypium anomalum* distributed in Angola and Namibia and the closely related if not conspecific G. *capitis-viridis* found in the Cape Verde Islands, presumably having reached these islands via sea currents, inasmuch as the basically inland distribution of G. *anomalum* reaches the coast in southern Angola.

What do these examples tell us? They do not tell us that the comose seeds of the Gossypieae are an adaptation to saltwater dispersal. I reach this conclusion from a consideration of the ecological distribution of these plants: none of them (with two exceptions) are strand plants. (The two exceptions, to be discussed presently, are *Thespesia populnea* and the tetraploid species of *Gossypium*.) They are basically inland plants, occurring often in arid habitats. Where these distributions chance to extend to a coastline (as is the case with *Gossypium anomalum*, noted above), their seeds may be picked up by the sea and transported to distant places. That is, in teleological phrases, they are capable of taking advantage of such a situation opportunistically and of using their comose seeds (which aid in flotation) for dispersal to new sites. Their seed hairs are thus a *preadaptation* to this mode of dispersal, as are their impermeable seed coats. Other species that reach coastal sites but lack seed hairs (such as *Gossypium cunninghamii* and G. *populifolium*) do not show this dispersal potential. The exception to this statement is the closely related species-pair G. *davidsonii* and G. *klotzschianum*, which occur in Baja California and the Galápagos Islands, respectively. The mode of dispersal that gave rise to this disjunction and divergence remains an unsolved puzzle. Both species have essentially hairless seeds.

The most notable example of evolutionary opportunism and preadaptation, however, is to be found in the tetraploid species of *Gossypium*. The origin of these tetraploids has been expounded elsewhere, especially in cytogenetic terms, and need not detain us, but the ecological aspects of the story deserve elaboration and emphasis.

Reference has already been made to the distribution of G. *hirsutum* to far-flung corners of the Pacific Ocean and to the evident relationship of this distributional pattern to ocean current patterns. Opposing views have been advanced to interpret these patterns, the disagreement concerning whether the plants achieved their present distribution independently of man by virtue of their own dispersal devices or whether they were spread by man as cultigens and subsequently "escaped" to their present habitats. The materials relevant to the argument include not only the "wild" cottons of the Pacific (from Socorro Island, the Marquesas, Samoa, Wake Island, and so on), but also those from the coasts and islands of Florida, Yucatán, Curaçao, Venezuela, and so on, on the Atlantic side of Middle America as far north as Tamaulipas.

The argument heretofore has involved data on the experimental capability of seeds for saltwater transport and survival, observations on the morphology of the plants themselves and the significance of the morphological characters for "wild" vs. "escaped" status (which inevitably remains a matter of interpretation), citations of relevant historical sources concerning the early voyages to the Pacific in support of the thesis of dispersal by man (which is inconclusive at best because of the limitations of the evidence), and observations on the ecology of the plants and of the habitats they occupy. The last argument appears to have the greatest prospect of elucidating the situation and will be elaborated on here.

It has previously been noted that the origin of the tetraploid species of *Gossypium*, which is the only case of polyploidy known in the tribe, was intimately tied to the invasion of a new ecological niche. The diploid species of the genus are found in inland, often arid, habitats. Even where certain of them occur on islands, they are from the interior and show no indication of being strand plants. The tetraploid species *G. hirsutum*, *G. barbadense*, *G. darwinii*, and *G. tomentosum*, at least those representatives that may be suspected of being "wild," are typically strand plants, occurring directly on the beaches in littoral vegetation or back from the beaches in littoral-derived vegetation.

This is a distinctly different ecological niche that these tetraploids occupy, and it may be regarded as a development that has occurred relatively recently in the evolution of the genus. The tetraploids are thought possibly to have originated in the Pleistocene (Fryxell, 1965b; Phillips, 1963). It seems significant that the littoral environment was invaded by these new amphidiploids at a time when both sea level and shorelines were fluctuating as a result of the repeated retreat and advance of Pleistocene glaciation. In this sense the mobile shorelines were "disturbed habitats" of the kind known to be fertile ground for evolutionary change. Moreover, the fluctuating shorelines maximized the opportunity for seed transport via ocean currents for seeds capable of such transport. Evidently the tetraploid Gossypieae used this opportunity.

If we adopt this view, then, we must focus our attention on these littoral plants and set aside the many other cultivated, semi-cultivated, and derived types of cotton (especially of *G. hirsutum*) that tend to obscure the picture of early evolution and of seed dispersal. Whether these "wild" cottons that are distributed so widely in the Pacific and the Caribbean are escapes from earlier cultivation or are "primitively wild" (that is, essentially uninfluenced by man) may be argued on various grounds. Previous arguments have been based on morphology and history, but the literature is witness to the fact that these data are subject to varying interpretations. Even the argument based on distribution, citing examples of places where the wild cotton is found but where agricultural man has had little or no influence (for example, Socorro Island), is not completely convincing, since good examples are so few and are never entirely clear-cut because so few places are beyond the influence of man. Thus, the question has long remained a matter of disputed opinion with little solid evidence to support the differing viewpoints.

New evidence, however, was put forth by Sauer (1967, pp. 24, 29–30) that offers substantial support to the view that these littoral cottons are primitively wild rather than derived. Sauer's observations of the wild cotton of the northern coast of Yucatán are part of an analysis of seashore vegetation of the entire Mexican Gulf Coast; they indicate that the cotton is "common and often dominant for a distance of over 100 km between Sisal and Dzilam de Bravo" and that it is a member of "a complex vegetation type occupying a coherent and extensive area with natural and edaphic and climatic boundaries." Most significantly, it is negatively associated with human settlement, being "absent only near habitations and where coconuts have been planted." These facts, especially the negative association with man, argue strongly for the view that this cotton is a truly wild plant with no history of domestication by man and, by extension, that this characteristic is true of littoral cottons generally. The littoral wild cottons may be regarded as relics of the early (preagricultural or even prehuman) amphidiploid Gossypia.

This view, then, leads us back to considering the evolutionary opportunism of the tetraploid species of *Gossypium* in terms of their seed dispersal mechanisms. Transport by sea, locally or over long distance, was evidently a principal means of dispersal—at least before man came along and recognized the utility of the fibers and proceeded to further disperse the plants in a second wave of evolutionary opportunism. The characteristics that made this dispersal by sea possible, including the impermeable seed coat, the relatively long-lived embryo, and the highly developed seed hairs, were not developed in response to selection for this mode of dispersal. Rather, they are much more ancient, as we have seen, and were present in the newly formed tetraploids as preadaptations that permitted the plants to use this mode of dispersal successfully and to invade the new (for the genus) ecological niche. Other factors that combined with this preadaptation were the capacity of the plants to grow in a relatively saline environment involving exposure to salt spray, as yet unknown historical considerations concerning the place and circumstances under which the tetraploidy originated, and the opportunity for invading the littoral environment (as a "disturbed habitat") provided by the mobility of shorelines during the Pleistocene.

One final example of dispersal in the Gossypieae remains to be considered—the remarkable case of *Thespesia populnea*. This small tree occurs throughout the tropics and subtropics as a strand plant. It may be found on the shores of all the principal landmasses and on most of the world's tropical islands, even the most remote. Its dispersal potential makes it truly pantropical. It has achieved a modest extension of this distribution as a cultigen, insofar as it is grown as a street tree in such places as southern India, Florida, and Australia, where it is valued for both its handsome foliage and its showy flowers. But its natural distribution has been achieved through the exceptional dispersal capacities of its fruits, which are the unit propagules. The fruit is an oblate indehiscent capsule with five cells. The carpel walls are leathery or semiwoody and evidently quite resistant to saltwater. The interior of the fruit is filled with spongy tissue in which the seeds are embedded. The entire fruit is about 3 cm in diameter; the individual seeds are about 1 cm long. The spongy tissue, together with the waterproof exterior, renders the fruits quite buoyant. They retain this buoyancy for considerable periods and thus can be, and are, transported over great distances by ocean currents. Eventually exposure causes the decomposition of the carpel walls, freeing the seeds for germination.

Presumably this decomposition is hastened by the drying out of the fruit after it is washed ashore and by exposure to the heat of the sun on the beach. The establishment of seedlings and the growth of the trees is confined essentially to the open beach, where exposure to salt spray and brackish water limit the plant population to those species that can tolerate these extremities.

The strand habitat has thus provided the principal stage for the dispersal of *Thespesia populnea* and for *Gossypium hirsutum*, both of which are examples of "extreme outpost plants" (Sauer, 1967), that is, those which grow immediately adjacent to the sea. It is the extreme outpost plants that are best adapted to sea dispersal by flotation, as compared to those strand plants that occur farther inland, which are often less well adapted for or completely incapable of such dispersal. I have referred to the strand habitat as a "disturbed habitat" in the sense that shorelines have a certain mobility, and, given that mobility, the narrow transect of the habitat is quite vulnerable to destruction. Indeed, the microhabitat is highly unstable, and the plants that inhabit it may be regarded as pioneer vegetation, in terms of Clementsian succession. But Sauer has pointed out that this very instability is in itself highly stable; that the littoral habitat, with its sharp gradations, narrow zones, and vulnerability to destruction, is a very ancient habitat; that the "pioneers" are simultaneously "old residents"; and that the habitat survives in the long run whatever short-term destructions it may suffer or be vulnerable to. It is in such a stable-unstable habitat that *Thespesia populnea* is so firmly established and into which the tetraploid Gossypia have (in evolutionary terms) recently entered.

Thespesia populneoides is a near relative of *T. populnea*. It shares both a similar morphology and a preference for the littoral habitat; it does not share the pantropical distribution, however, being confined to localities around the Indian Ocean. Since *T. populneoides* has buoyant fruits that are only imperfectly dehiscent and occurs in the littoral environment, one would expect it to be as widely distributed as *T. populnea;* yet it is not. Perhaps the partial dehiscence of the fruits has proved limiting. Or perhaps the reason is to be found in the breeding system, if *T. populneoides* is indeed self-incompatible as preliminary observations suggest. If this is true, then the much wider distribution of the self-compatible *T. populnea* is another example of the operation of Baker's Law, which states that self-incompatible plants can achieve long-distance dispersal only by means of self-compatible variants. Such a change in breeding system, with its consequent effects on distribution, may have been part of the impetus behind the evolutionary divergence of these two species.

5

Ecological Limitations on Spread

*Adaptation cannot but be universal among organisms, and
every organism cannot be other than a bundle of adaptations,
more or less detailed and efficient, coordinated in greater or
lesser degree.*

                                                    J. Huxley, 1943

Frost and the Tropical Imperative

MEMBERS of the Gossypieae are essentially excluded from the temperate zones and are confined to the tropics and subtropics. To better
understand this distributional constraint, we should examine those exceptional cases of species that have "escaped" the tropics, if only to a
limited extent. Perhaps the most notable example is *Gossypium thurberi*, which grows in the Sonoran Desert as far north as southern
Arizona. It extends to 33° N latitude and to elevations as high as fifteen
hundred meters—rarely even higher. The climate of these regions is
clearly temperate, with severe freezes every winter.

The species has evolved a group of adaptations that permit it to
survive in a temperate climate. The principal adaptations are (1) a
deciduous habit, whereby the plant enters into a dormant phase before
the advent of frost, (2) an ability of the dormant plant to withstand low
temperature to a degree unusual for the genus, indeed, for the tribe,
and (3) a precisely timed flowering and fruiting cycle, with very rapid
fruit maturation that enables seed production to be completed before
the dormant period commences and frost occurs. Each of these adaptations merits separate comment.

Although many of the species of *Gossypium* are essentially evergreen, several are deciduous and regularly pass through a dormant
stage. In these cases, however, the dormancy is not related to a warm-
season–cold-season cycle, but to a wet-season–dry-season cycle. The
most notable examples of this pattern are the species of *Gossypium*
subsect. *Erioxylum*. These species grow vegetatively during the wet
season (July–September). By November they begin to lose their

leaves. In December and January (when they are leafless) flowering occurs, and fruit maturation follows while the plants are still in a leafless condition. The cycle is completed by March or April. Then follows a period of dormancy, which lasts until the advent of the rains sees the vegetative cycle begin again.

The seasonal cycle of *G. thurberi* is evidently an adaptation of this wet-season–dry-season cycle, because the renewal of vegetative growth in this species is clearly tied to the advent of summer rains. In southern Arizona these rains typically begin in July, and the regrowth of *G. thurberi* also typically begins then. In exceptional years, however, when these rains begin earlier or when there is an unusually high carry-over of soil moisture from a wetter-than-average winter, regrowth may begin as early as April, when the season first becomes warm enough for the winter dormancy to end. In dry years it may begin later than July. The *initiation* of this winter dormancy, however, is not a matter of declining moisture supply but of lowering temperature and/or changing day length. Regardless of the precise mechanism of initiation (which is left to the physiologists to elucidate), it is clear that the triggering stimulus for the initiation of dormancy is a seasonal one relating to the temperate climate and not one relating to the desert habitat. It is this adaptational pattern that is unusual for *G. thurberi*.

In no case (apart from the cultigens; see p. 169) have members of the Gossypieae adopted the annual habit as a means of surviving in a temperate climate. Only in the genus *Cienfuegosia* do we find the reduced herbaceous plant habit that is typical of annuals. All of these, however, are perennial herbs that live over by means of a perennial woody rootstock. None of these species has extended its range beyond the subtropics, presumably because in no case has a species of *Cienfuegosia* evolved a protoplast capable of withstanding frost.

It is this ability to withstand freezing temperatures that is the second facet of the adaptational syndrome of *G. thurberi* that relates to its temperate distribution. It does not stand entirely alone in this regard, however, because the Australian *Gossypium sturtianum* also has a degree of frost tolerance. *G. sturtianum* also occurs as far beyond the tropics and subtropics as the thirty-third parallel, although it does not occur at as high elevations as does *G. thurberi*. *G. sturtianum* is essentially evergreen and does not couple its ability to withstand frost (while in full leaf) with a dormancy pattern or any other adaptational feature. Moreover, its ability to withstand frost is limited to a few degrees below freezing and is not nearly as extreme as the ability of *G. thurberi* to survive hard freezes when in the dormant condition. Again, we find *G. thurberi* standing apart as an unusual representative of the Gossypieae.

It might be noted parenthetically that while the pattern evolved by *G. thurberi* has greater interest in evolutionary terms, the frost-tolerant characteristic of *G. sturtianum* may deserve greater attention from the point of view of applied plant breeding. The fact that the frost-tolerant characteristic of this species is not coupled with a dormant stage makes it of greater applicability to a crop plant. This is especially true of cotton, which is a tropical plant but which, as an annual cultigen, is being pushed further and further into the temperate zone. Work in this direction has been begun by Muramoto (1969) in a unique breeding experiment at the hexaploid level that had other objectives but in which the greater cold tolerance of the *G. sturtianum* parent was incidentally brought across into the synthesized polyploid.

The third aspect of the adaptational pattern of *G. thurberi* is its reproductive cycle, which has necessarily been shifted in response to the imposition of a winter dormancy period. The Mexican species of *Gossypium* subgen. *Houzingenia* (of which *G. thurberi* is a part) typically flower in months that in a north temperate zone would be late autumn or winter. The nearest relative of *G. thurberi*, the subtropical *G. trilobum*, flowers in the months of October and November in southern Mexico and matures its fruits during subsequent months. *G. thurberi* in Arizona, on the other hand, begins flowering in mid-September and ceases by mid-October. Fruit maturation begins in early November at about the same time that symptoms of leaf senescence first herald the onset of dormancy. Thus, by flowering early and rapidly, and by maturing its fruits in a relatively short period of time, *G. thurberi* completes its reproductive cycle before the onset of dormancy—and, incidentally, before many of its relatives farther south have even begun their reproductive cycles.

Yet the striking observation to be made in the evolutionary context is that *G. thurberi* has achieved this adaptation by stretching and straining at an existing pattern, not by breaking out of the pattern to something wholly new. *G. thurberi* has pushed back its reproductive cycle to a degree sufficient to enable it to fit the seasonal cycle of a temperate climate. But it is still dependent upon the triggering effect of the advancing season (a photoperiodic response, to use that sometimes misapplied term) to initiate flowering; it has not become day neutral in a way that would permit summer flowering and further penetration of the temperate zone. The adaptational pattern shown by G. thurberi exemplifies the principle of "adaptive modification along the lines of least resistance" as put forward by Stebbins (1974, pp. 31ff.).

Many species of Cienfuegosia have the summer flowering habit but, as we have seen, lack the cold tolerance that would enable them to extend beyond the subtropics. With the exception of Cienfuegosia and a few species of Gossypium (for example, G. bickii), the winter flowering habit (short-day photoperiodic response, if you will) is firmly entrenched in the heritage of the Gossypieae. The inability of G. thurberi to break out of this pattern emphasizes its deep-seatedness or "evolutionary canalization" (Stebbins, 1974).

Alternatively, one might speculate that G. thurberi has not been able to discard the photoperiodic basis for the initiation of winter dormancy. The balance of selection pressures has given rise to the present pattern. It should be possible to test this matter experimentally, because strains of G. thurberi are known in cultivation that are day-neutral in their flowering response. However, I am not aware that any test has been made of the response of these strains to the changing conditions of the advancing season of their natural environment, in terms of their entering into the dormancy stage.

The distributional pattern shown by these plants, together with the physiological basis for it, which as a whole I term "the tropical imperative," may be seen as a basis for rejecting as untenable the view advanced by Stebbins (1947, 1959) that Gossypium migrated from the Old World to the New World via the Bering land bridge in early Tertiary times. Not only were the minimum temperatures of these latitudes extreme for members of the Gossypieae, even in the milder climates of the Eocene, but the strongly canalized reproductive patterns that are characteristic of Gossypium would be completely dis-functional at the latitude of Beringia. The most extreme example available in Gossypium, G. thurberi, has been able to adjust its reproductive pattern only to the limited extent of permitting it to occupy latitudes as far north as 33° but no farther. To propose that cottons related to the present-day G. arboreum and G. herbaceum expanded through Siberia to the Arctic Circle and thence into the New World is not realistic.

The Puzzle of "Photoperiodism"

THE title of this section deserves a word of comment. In it I use the word *puzzle* advisedly, because I confess myself unable to resolve some of the problems that are raised by the phenomena to be discussed. I put quotation marks around the word *"photoperiodism"* because I consider it to be a misleading and unsatisfactory term in description of the flowering patterns to which the term is commonly applied. But a term is needed, and "photoperiodism" is most readily available.

Anyone who has tried to grow any of the wild tropical representatives of *Gossypium* in a temperate latitude is familiar with the phenomenon that I am talking about. When grown outdoors in the summer growing season of a temperate latitude, the plants remain completely vegetative. They can generally be induced to initiate and develop flowers only by transferring them into a greenhouse for the winter months. Since the most obvious difference between the summer and winter environmental conditions (in the temperate latitude) is that of day length, this flowering pattern is commonly interpreted as a manifestation of photoperiodism, and the plants in question are regarded as being short-day flowering plants. Experimental data dealing with day-length manipulation generally support this view.

I do not wish to reject the notion that this phenomenon is indeed an aspect of photoperiodism but only to suggest that it is merely *one* aspect of a complex situation. It is difficult to divorce variations in photoperiod from other climatic factors that vary seasonally with day length, even under experimental conditions. To ascribe the response to a single factor, selected out of the total environmental milieu, is probably not accurate, especially since it is already known that night temperature is another factor that influences flowering in these plants. Moreover, certain species cannot be induced to flower with simple manipulations of day length, indicating that other factors are at work.

Consider this admittedly extreme example. *Gossypium tomentosum* is very difficult to induce to flower in many habitats other than its native Hawaii. It can be more reliably induced to flower by girdling or by inverting the phloem than by any manipulations of day length or night temperatures. How meaningful is it to call this a "photoperiodic" flowering response? Cool temperatures seem to be at least as important as day length.

Floral induction and floral development are apparently separate phenomena, a fact that must be taken into account in resolving the problems of interpreting these patterns. Under certain conditions plants can be stimulated to *initiate* flowering, but the buds that are initiated fail to develop, and they abscise at an early stage, indicating that additional stimuli for floral development are also required.

Often plants taken out of their normal environment (often to distant latitudes) behave in a most erratic fashion with respect to flowering patterns. For example, I have grown the Australian G. *sturtianum* in the open in Arizona, where it remained completely vegetative for several years, growing vigorously in the summer and often dying back to the ground as the result of winter freezes. One year a mild winter permitted it to remain in leaf throughout the year, as a result of which floral induction took place and heavy flowering occurred in late winter. Once floral induction had taken place, the plant remained more or less in flower (depending on fruit load) *throughout the year*.

Another example concerns *Gossypium thurberi* from Arizona and Sonora. As described elsewhere (p. 152), this species has a delicately timed flowering and fruiting pattern that is very closely related to the seasonal cycle. It is thus a "photoperiodic" species. Yet when taken out of its natural habitat, it behaves oddly. In south-central Texas (which, although it is in essentially the same latitude as Arizona and Sonora, has a greatly different climatic pattern, differing in having more moisture and less insolation), G. *thurberi* has an erratic flowering pattern. Whether in the greenhouse or in the field, it may come into flower at almost any time or fail to do so. It certainly lacks the highly predictable flowering pattern that it exhibits in nature in the Sonoran desert. Clearly, there are factors of the environment other than day length that are controlling its flowering response in "foreign" environments.

In yet another example, the species of *Gossypium* subsect. *Erioxylum* (G. *aridum*, G. *lobatum*, and G. *laxum*) are also highly seasonal in their flowering; yet photoperiod is evidently not the controlling factor in their pattern. These species grow in habitats that are characterized by a severe dry season alternating with a wet season. The beginning of the wet season (June–July) initiates vegetative growth. With the coming of the dry season (September–October), growth slows and ceases. Leaf senescence is followed by leaf abscission (November–December). Flowering commences only when the trees are essentially leafless, and fruiting is completed (January–February) only just before the severity of the dry season enforces the complete dormancy of the plants. Limited experimental evidence indicates that dryness, not day length, controls this flowering pattern. For these and other reasons there are many qualifications that need to be made in calling the distinctive seasonal flowering patterns of certain of these species "photoperiodic."

How widespread is the phenomenon of "photoperiodism" in the Gossypieae, however it is interpreted? That question is only partially answerable, because it is not possible to fully understand the phenomenon without cultivating the plants in "foreign" environments, and many of the Gossypieae have not yet been brought into such experimental culture. Not surprisingly, the best information comes from the genus *Gossypium*, of which most of the species have been studied experimentally. However one interprets the meaning of the term *photoperiodism*, the majority of the species of *Gossypium* exhibit such a flowering pattern. The principal exceptions are the cultigens (which have undergone intensive selection to "break" this flowering pattern) and the Australian *G. australe* and *G. bickii*, which seem to flower normally in temperate-latitude summers.

In *Cienfuegosia* I have grown about half of the species in artificial cultivation. Of these, only one, *C. affinis*, shows an unmistakable photoperiodic response, flowering only in the short days of winter in greenhouse cultivation. Other species flower somewhat erratically (for example, *C. rosei* and *C. digitata*), but most flower freely.

In *Hampea* the flowering pattern is different. Flowering is definitely seasonal, occurring principally from August to October for all species so far as is known, but it does not seem to be controlled in the same way that the pattern for *Gossypium* is. Limited data from growing species of *Hampea* outside of the tropics indicate that they retain the same pattern. The controlling factor in floral induction is evidently less subject to upset (or manipulation) in *Hampea* than it is in *Gossypium*.

Certain species of *Thespesia* seem to be "photoperiodic" in the same manner as *Gossypium*, but less evidence is available. *Thespesia garckeana*, *T. thespesioides*, and *T. beatensis* are definitely short-day plants in greenhouse cultivation. It is difficult to bring about floral induction in *T. danis* and *T. grandiflora* in my greenhouse experience. *T. populnea* is erratic in its flowering response. *T. lampas* is seasonal to a degree in its flowering pattern but tends to flower rather more freely than other *Thespesias*, perhaps as a result of its history as a cultigen.

Adequate evidence is not available for the other genera of the

Gossypieae except to say that limited experience with cultivating *Kokia* in the greenhouse failed to produce any flowering. My experience in cultivating *Lebronnecia* in the greenhouse is too limited to draw any conclusions except that the timing of flowering seems to be in accord with that of the plant in its native habitat. *Cephalohibiscus* is not known in experimental cultivation. Its natural flowering period (in the Southern Hemisphere) is in July and August, and fruiting follows in October and November. Although this pattern is deduced from rather limited data from collectors' notes, it appears to be a flowering pattern that is definitely seasonal. Whether this seasonality would be stable in other environments is, of course, not known.

The "puzzle" of photoperiodism does not concern those species (for example, *Hampea* and probably *Cephalohibiscus* and *Lebronnecia*) that show a seasonal flowering pattern in their native tropical environs, even though questions remain concerning the mechanisms of control of these seasonal patterns. These examples are not a puzzle, because the patterns are clearly adaptational, even though the bases for the adaptation may not be understood.

The "puzzle" concerns those species that are not seasonal in their flowering patterns in their native tropical habitats but become "photoperiodic" when transferred to temperate latitudes. Days do not vary in length significantly through the annual cycle in the tropics and are never far removed from the equinoctial state. Why should plants whose antecedents are tropical respond so rigidly to a day-length regime that is foreign to them? The response is firmly established in the germplasm of many of the species (and even genera) of the tribe Gossypieae but is by no means confined to that group. It is a phenomenon of some generality. Presumably it is fixed in the germplasm by selection for adequate adaptational reasons. But what are those reasons? Therein lies the puzzle.

Evidently these tropical plants are adapted to the approximately equinoctial conditions of the tropics, under which conditions they are able not only to grow but also to initiate and develop flowers. Outside the tropics (barring freezing temperatures) they are able to grow quite well, but floral initiation and development is more refractory. The plants are unadapted to the temperate zones, at least in terms of flowering in the usual summer growing season. Their tropical adaptational pattern requires some minimum day length (or maximum night length) to trigger floral initiation and development—a requirement that is met by the short days of the temperate winter but not by the long days of the temperate summer.

Thus, the response is understandable, but the adaptational basis for it remains to be explained, especially in view of the depth of its entrenchment in the genotypes of these plants. "Photoperiodism," in the superficial sense of differential response to varying day length, makes no evolutionary sense when it occurs in tropical plants that are exposed to a relatively constant day length. Is it possible that this phenomenon is related to the inferred austral origin of the tribe? that "photoperiodism" is a canalized relict of the earlier occupancy of milder climates in higher latitudes of the Southern Hemisphere by the progenitors of the tribe?

The problem is regarded as an important one because it may yield information on the evolution of temperate plants from tropical progenitors. If the "photoperiodic" flowering response described here is indeed a relatively general phenomenon, then an understanding of how it may (or must) be modified for the successful invasion of the temperate climates of higher latitudes will contribute significantly to our knowledge of angiosperm evolution. An understanding of the basic pattern is the first step toward a knowledge of the broader problem.

6

The Influence of Man

It is the simplest questions which are the hardest to answer. . . .
One of these simplest of simple questions has been hiding
around the corner from us. . . . What is a wild plant? Why, a
plant that is growing wild, to be sure. But how does one know
when it is growing wild?

Edgar Anderson, 1969

How has man influenced the plants of the tribe Gossypieae?—in various ways and to varying degrees. The tribe Gossypieae, in fact, is of special interest because it proves to be a microcosm of the entire plant world in terms of the influences it has received from the activities of man. Some of these influences have been negative, resulting in restricted distributions or even extinction of certain taxa. Other influences have been more positive, changing the direction of evolution of certain groups or assisting in their spread to new areas.

Extinction and Imminent Extinction

ON the negative side we may note the extinction of certain species of *Kokia* and the threatened and possibly imminent extinction of the entire genus from the Hawaiian Archipelago, where it is endemic. *K. drynarioides*, though possibly extinct in nature, persists on a small scale in cultivation as a park and street tree in Hawaii. *K. cookei* is also extinct in nature but still survives in a few places where it has been planted. *K. lanceolata* is certainly extinct. Only *K. kauaiensis* survives in nature on the island of Kauai, but its populations are severely limited. It is not known to be in cultivation, but it deserves to be, because its flowers are much larger than those of the other species.

The adverse human influence on *Kokia* has largely been indirect but none the less devastating for all of that. The species have become depleted or extinct in part because of the direct activities of urbanization and land clearing, although the native Hawaiians in earlier times stripped the bark from *Kokia* trees for a red dye (containing gossypol?)

used as a preservative for their fishnets. But the more pervasive influence has been through the destruction of the habitat and by the competition of aggressive adventive plants introduced by man during the last two centuries. Island endemics are often poorly equipped for competition (cf. Carlquist, 1974); the weedy fellow-travelers of man are often especially aggressive competitors. The contest is uneven, and *Kokia* is one of the losers. This loss is offset to a degree by the protection offered to *Kokia* in its role as a cultigen. This cultivation, however, is on a very limited scale, and the future for *Kokia* looks bleak.

Interestingly, the wild cotton, *Gossypium tomentosum*, also a Hawaiian endemic, does not seem to suffer from this same competitive disadvantage when challenged by adventives. This fact may be used to argue for its relatively more recent arrival in Hawaii. *G. tomentosum* is reported to be suffering inroads, however, from the more direct human activities of construction and urbanization.

Other taxa of the Gossypieae, most notably *Lebronnecia kokioides* from the Marquesas Islands and *Thespesia beatensis* from the Caribbean, also appear to be on the verge of extinction. In the former case, at least, man's introduction of alien competitors seems to be a major reason for the decline of this monotypic genus.

Other relatively narrow endemics, however, do not seem to be threatened by man, at least at present. One might cite the *Thespesias* of New Guinea, the *Cienfuegosias* of Arabia and Somalia, and some of the *Gossypiums* of northern Australia, for example.

Genetic Challenges

THERE are a few cases of relatively minor human influences on the plants in question. For example, human agricultural activities in Venezuela resulted in the introduction of the notorious insect pest the boll weevil (*Anthonomus grandis* Boh.) to that country as a part of its cotton agriculture. Although the host range of this insect is quite narrow, being essentially confined to plants of the tribe Gossypieae (Cross et al., 1975), it had no difficulty establishing itself on *Cienfuegosia affinis*, which is indigenous to Venezuela. That plant is thus faced with a new and severe insect pest. The introduction of the weevil is known to have occurred as recently as 1949; therefore, it is as yet too early to evaluate the plant's response to this evolutionary challenge.

Another challenge resulting from human activities has been given to *Gossypium darwinii*, endemic to the Galápagos Archipelago. This challenge has come from the introduction in historical times of the closely related cultigen *G. barbadense*, at least to certain of the inhabited islands. The two species (or varieties, depending on taxonomic opinion) are sufficiently closely related to interbreed freely, and the challenge has come from a swamping of the genotype. There is unquestionably significant introgression occurring, the one plant contributing the greater vigor of the cultigen, the other the capacity to survive naturally in the exacting environment of those arid islands. How natural selection will deal with this interaction remains to be seen. A similar situation is occurring on a more limited, localized scale in Hawaii with *G. tomentosum*.

The Polynesian form of *Gossypium hirsutum* is indigenous to a number of islands of the Pacific. In the Marquesas Islands (and to some extent elsewhere) it, too, is receiving a genetic challenge as a result of man's agricultural activities. In the past century there was considerable introduction and cultivation of cotton from South America (*G. barbadense*). The Polynesian cotton industry has long ago died, but the introduced plants have persisted in a feral state. In some cases they have come into immediate contact with the indigenous *G. hirsutum*, and hybrid and introgressed forms may be found. The effect in this case, however, is relatively minor, since the two species have a significant genetic barrier between them that minimizes the extent of gene exchange. Although they form fertile hybrids freely, extensive genetic breakdown and sterility occur in subsequent generations, so only very limited introgression can occur.

Friends: Plants That Are Used but Not Cultivated

A more direct influence of man occurs for those species that are either used or cultivated by him. At the very lowest "use" level—that of plants that are not cultivated but parts of which are gathered and used —we may cite many of the species of *Hampea*. Strips of bark from these trees are used throughout Middle America to make a crude rope for tying bundles and for similar uses. The trees are often known by the vernacular name "*majagua*," which is appearently a generic term for all trees having this same use, including several other genera not closely related to *Hampea*. One result is that when land is cleared for cultivation, the *Hampea* trees are often spared because of their known usefulness.

Only slightly above this level of use is the position of

Cephalohibiscus peekelii, which occurs in New Guinea and several of the Solomon Islands. These trees are useful to the indigenous peoples of the area in two ways: the timber is used in the construction of dwellings, and the seed hairs are used for the same purposes as are those of kapok. According to Ulbrich (1935), the trees are indigenous to New Guinea but have been spread by man to other areas to a limited extent because of the usefulness of the seed hairs.

Companions: Plants That Are Cultivated but Not Selected

A few species are planted as ornamentals but have received little or no selection, only more widespread distribution, by man. These include *Kokia drynarioides*, planted as a street and park tree on a limited scale in Hawaii; *Thespesia populnea*, used similarly in India, Florida, Australia, and probably elsewhere; *Thespesia grandiflora*, planted as a specimen tree in the Caribbean, Central America, and Hawaii; *Thespesia lampas*, grown as a specimen shrub in various parts of the tropics; and *Gossypium sturtianum*, grown as a specimen shrub in Australia.

Slaves: Plants That Have Undergone Intensive Selection (*Gossypium*)

WITHOUT question, the most significant effect of man on these plants has been on the cultivated species of *Gossypium*, which have been the basis for a large-scale economic enterprise, the cotton industry. The effect has been enormous on the four cultivated species; it has been essentially nil on the wild species of *Gossypium* or on the other genera of the Gossypieae. The cultivated species have been virtually transformed at the hand of man. Some of them have been spread around the globe; the origins of some have been lost from sight. Let us attempt to review some of the highlights of this story.

First we must review the origin of tetraploidy in *Gossypium* in order to grasp the entire picture. Extensive cytogenetic study of *Gossypium* has shown the existence of certain genome groups—groups of species showing a high degree of homologous chromosome pairing in hybrids within the group but a low degree of pairing in hybrids between different groups. The taxonomic arrangement presented in chapter 2 in part reflects this cytogenetic information. These genome groups, for convenience, have been identified by cytogeneticists with capital letters A through G. The question arises, What is the genome constitution of the tetraploids?

It has been well established on the basis of work done by many geneticists and cytologists, but especially on the basis of pioneering work done by J. O. Beasley (for example, Beasley, 1940*a*, 1940*b*), that the tetraploid cottons have an AD constitution (Fig. 53). That is to say, the tetraploids combine genetic materials of the A genome (of the Old World cultivated diploids) with genetic materials of the D genome (of the New World wild diploids). It is firmly established that the tetraploids originated from the hybridization of two such dissimilar plants accompanied or followed by the doubling of the chromosome number to produce an amphidiploid. It is *not* firmly established how, when, or where the tetraploidy originated, and indeed, these questions have been the subject of endless discussions and only limited agreement.

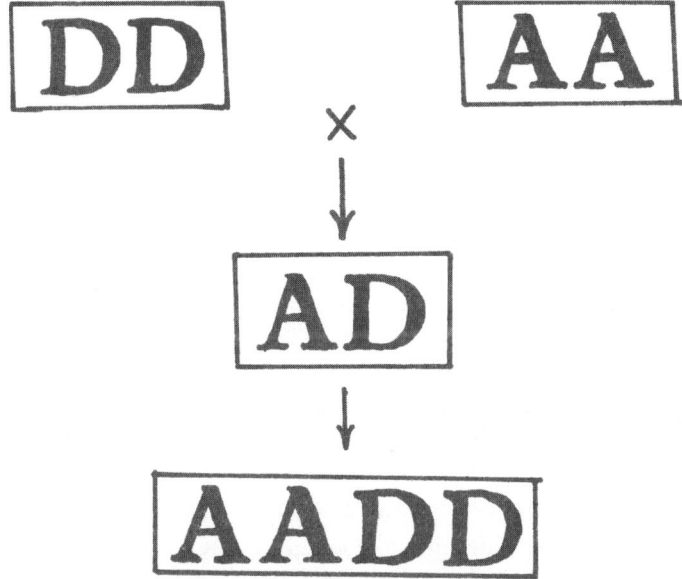

Fig. 53. Diagram of the origin of tetraploidy in *Gossypium*. A = haploid chromosome complement of diploid Old World species; D = haploid chromosome complement of diploid New World species.

It is not my purpose to review these inconclusive arguments. Let us simply accept that the tetraploids did, somehow, originate, and that they evidently originated and certainly became established somewhere in the New World tropics. The time of origin of the tetraploids has been placed by different botanists at various times from the Cretaceous to the Recent. There is little more than indirect evidence to bring to bear on the question, but a modern consensus would probably place the origin of the tetraploids in the Pleistocene. It should be emphasized that these conclusions are based on such relatively intangible evidence as degree of divergence among the tetraploids compared with a knowledge of the diversity of the diploid species and coupled with inferences about rates of evolution. There is no fossil evidence. The archaeological record is of little help. We are guessing, and we know it.

Yet the question is of importance, because we wish to relate the origin of the tetraploids to their domestication and thus to the presence of man on the scene—or perhaps more accurately to the presence on the scene of man *and* of developing agriculture.

Although there is no general agreement on the subject, it seems clear to me that the tetraploid *Gossypiums* originated independently of man—whether or not they originated contemporaneously with man or were present on the scene before his arrival in the New World. I draw this conclusion from a consideration of the overall ecological picture. The following are some of the salient points:

1. There is great diversity among the tetraploids, involving differentiation into several species by even the most conservative views. This diversity implies existence over a period of time greater than the time available since the origin of agriculture in the New World, especially if we accept that divergence is slower at the tetraploid level because of the buffering effect of polyploidy.

2. The geographical spread of the tetraploids (Fig. 54), with *G. tomentosum* in Hawaii, *G. darwinii* in the Galápagos, *G. mustelinum* in northeastern Brazil, *G. hirsutum* and *G. lanceolatum* in Middle America, and *G. barbadense* in South America, similarly argues for the evolution and dispersal of the tetraploids over a longer time span than is available since the development of early agriculture.

3. A consideration of the morphology of extant wild (or apparently wild) tetraploid cottons persuades me that these are primitively wild and have not "escaped" and run wild from a previously domesticated state. Certain characters (for example, germinability of seeds, abundance of lint) occur in these wild forms in a state similar to what one might expect in a "raw" amphidiploid of the indicated parentage, perhaps of use to gatherers but showing no evidence of past selection for agricultural desirability.

4. The habitat of these wild tetraploids is distinctive; they occur mostly in littoral or littoral-derived habitats, which together with their

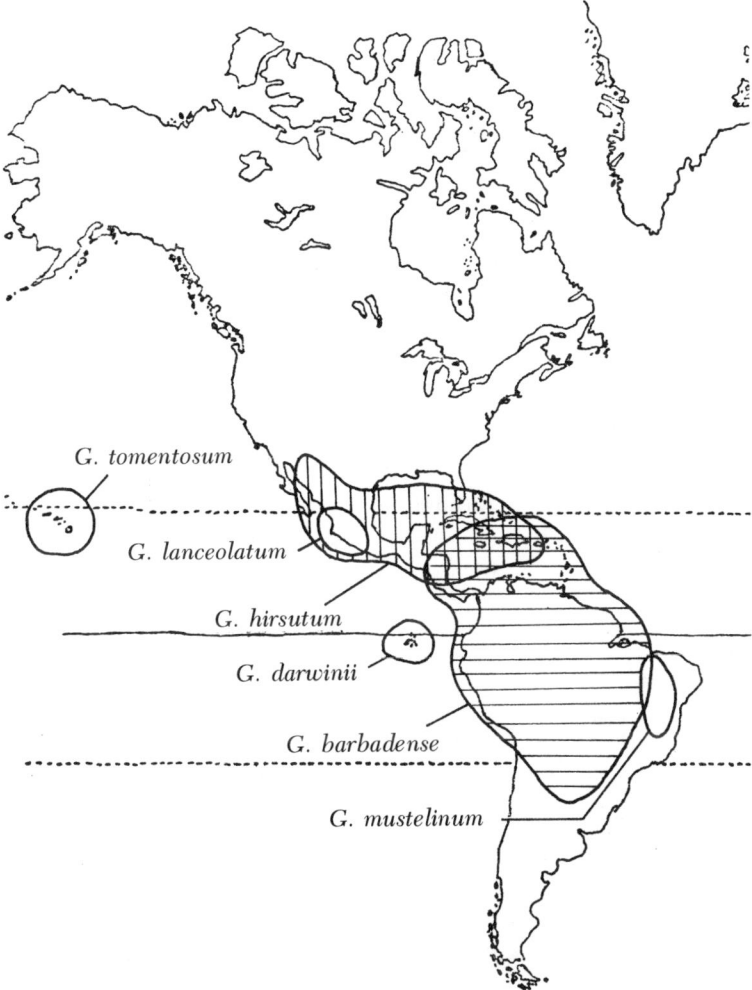

Fig. 54. Geographical distribution of the tetraploid species of *Gossypium*.

impermeable seed coats explains their wide dispersal on ocean currents. If these were feral cottons, run wild from previous domestication, they would not have this distinctive ecological distribution,
which is in virtual isolation from areas of agriculture.

All these points suggest to me that in the New World early agricultural man *found* tetraploid cottons and domesticated them; he did not
produce them. If this view is correct, the differentiation of the tetraploids, at least at the specific level, occurred before domestication, and

we can look to processes of natural selection to account for the divergence. Differentiation at the infraspecific level occurred in large measure under domestication, and we can view this pattern in agricultural and historical terms. Moreover, domestication of the New World tetraploids and domestication of the Old World diploids were independent occurrences of plant domestication.

The world's cultivated cottons may be divided broadly into two groups: the Old World diploids and the New World tetraploids. This simplified conception is due to the brilliant insight of the Soviet scientist Zaitzev, who published it in 1928. Modern treatments of *Gossypium* (Hutchinson, 1947b; Mauer, 1954; Fryxell 1969b) are all based on Zaitzev's contribution. Earlier works (most importantly, the monographs of Todaro [1877] and Watt [1907]) saw the immense welter of variability of the cultigens and, lacking the simplifying rubric of ploidy level and the insight gained from genetic experiments, failed to resolve this variability into an intelligible pattern. Certain recent studies (Prokhanov, 1947; Roberty, 1942, 1946, 1950) have failed to grasp Zaitzev's contribution and have presented atavistic (and useless) accounts of *Gossypium*. This history is reviewed in more detail in chapter 1.

The Old World (diploid) cultigens are divided into two groups in the Zaitzevian conception: the two Linnaean species *G. herbaceum* and *G. arboreum*. Each species is variable, encompassing both perennial forms in the tropics and annual types in more temperate latitudes. Only *G. herbaceum*, however, has an identifiable wild progenitor, the plant known as *G. herbaceum* var. *africanum*, which is indigenous to southern Africa in the Transvaal, Botswana, and Mozambique.

To be sure, there are those who claim var. *africanum* is not a wild progenitor but rather a wild derivative, escaped and become established in the natural vegetation from earlier cultivation. This, of course, is a difficult question to answer, but it seems clear that the characteristics of the plant are not those that would be expected in an escaped cultigen—dormant seed, sparse adherent lint, and the like. The plant is well established in the native vegetation of the area. As it exists today, it is a plant that would be attractive to gather and use for its seed hairs, requiring only that man have the realization that fibers are useful or the curiosity to experiment with a novel substance. Such curiosity might well have led to uses for the fibers, which in turn may have provided the impetus for the initial steps of domestication. From such a beginning, domestication and spread would readily follow. I cannot accept the argument that, since the fibers are useful, the plant cannot be a wild progenitor but must be an escaped cultigen. Such a view requires that primitive man *created* something useful from something that was not; it is not satisfied with considering that primitive man *found* something useful and improved and domesticated it. To say that any plant with usable fibers must be ruled out in our search for the progenitor of the cultivated cottons is to rule out any likelihood of success in the search.

Gossypium herbaceum, as a cultigen, is or was grown in parts of Africa, including Ethiopia and Arabia, and in the Levant countries as far as western India. Its cultivation is declining today because of the agricultural superiority of some of the New World cultigens that are being introduced and are displacing it. Where and when *G. herbaceum* was domesticated are certainly open questions, but that domestication need not have occurred in southern Africa, since trade routes to the Middle East along the east coast of Africa were open at an early date. Man not only distributed the plant over that part of the world, but also modified it considerably through selection over long millenia. Exactly how long the history of cultivation of cotton is is not yet clearly known because the archaeological evidence is still only scanty. The earliest archaeological finds from the Old World (Mohenjo Daro, ca. 3000 B.C.) nevertheless give evidence of an advanced cotton technology and thus indicate that the origins of cotton as a cultigen in the Old World must be much more ancient.

The wild var. *africanum* is a shrubby plant that branches rather widely. In early cultivation it probably did not differ greatly in plant habit, but modifications were induced in other characteristics of the plant. Probably the first characteristic affected was the fiber itself, and the ease with which it could be harvested and used. I would speculate that more readily detachable, finer fibers were the first modification, and more abundant fibers followed soon after. Selection for greater productivity is probably a relatively recent achievement. Intensive agriculture is not characteristic of the early stages of the domestication of plants. This is especially true of a plant used for purposes other than food, as is cotton. Such nonintensive cultivation may be seen today in various parts of the tropics where one or a few cotton plants are grown as perennials in dooryards and harvested sporadically. Larger-fruited types may be selected, but overall high productivity is not of great importance.

Gossypium herbaceum var. *africanum* may be regarded as a wild ancestor of the domesticated plants included in *G. herbaceum*—or at least as a model of such an ancestor. But what of *G. arboreum*, the other domesticated diploid species? It is sufficiently well differentiated from *G. herbaceum* that, in my opinion, the two could not have diverged in the few thousand years available since the beginnings of agriculture. Presumably these two species diverged at an earlier date and were subsequently brought into cultivation. Some recent archaeological finds (Chowdhury and Buth, 1971) have led to the suggestion that cotton (possibly *G. arboreum*) was cultivated in Nubia forty-five hundred years ago not for its fibers but for its nutritious seeds, which were valued as food for domestic animals. Others have proposed that *G. arboreum* is indigenous to India, where its greatest development in cultivation has occurred. Regardless of where or when the original domestication of *G. arboreum* took place, it evidently occurred *after* the differentiation of *G. arboreum* and *G. herbaceum*. Thus, we must conclude that they were domesticated independently.

In retrospect, then, it appears probable that the four domesticated species of cotton, the New World tetraploids *G. hirsutum* and *G. barbadense* and the Old World diploids *G. herbaceum* and *G. arboreum*, were all domesticated independently in four different sets of circumstances. These four occurrences were likely four different centers of developing agriculture and civilization. I will leave to others whether this supports or refutes diffusionist theories of cultural development, but in any case it appears that these ideas of domesticability of fiber-bearing plants were applied independently in four different parts of the world to four different botanical substrates. Moreover, it should be pointed out that the plants were not necessarily first domesticated for their fibers, but they may have had other uses.

The subsequent history of the evolution of the cultivated cottons is a remarkable story of parallel development. Although in prehistoric times these crops underwent domestication and evolution independent of each other, in more recent times all four have responded to similar selection pressures imposed by needs of civilization and an industrialized society on all of the domesticated cottons.

It is with the advent of industrial use of cotton that the industrial production of cotton (that is, field agriculture, as distinct from yard agriculture) has become important. It is thus only in times that are fairly recent in anthropological terms that cotton has become a field crop and has been subjected to selection for field production. Such selection has influenced such traits as (1) ease of germination (that is, loss of seed dormancy), (2) reduction of plant size to a more disciplined growth habit, (3) increase of productivity through increases in both the size and the number of fruits per plant and also through increases in the amount of fiber on each individual seed, and (4) development of the annual habit.

I have pointed out in an earlier work (Fryxell, 1965*b*) that the adoption of the annual habit was a major evolutionary step for *Gossypium*, that it was achieved as a result of human selection, and that it occurred independently four different times in a relatively short period of time. Annuals are unknown among the wild species of *Gossypium*. Consequently, because of the frost tenderness of the genus (indeed, of the entire tribe of Gossypieae) these plants have been unable to leave the tropics and enter temperate latitudes, except for the cultigens and two exceptional species, *G. thurberi* and *G. sturtianum*, which are discussed in chapter 5.

In the case of *G. herbaceum*, its cultivation was evidently pushed northward into temperate latitudes sometime before the thirteenth century after Christ but later than classical times. *G. hirsutum*, cultivated by the Pueblo Indians of the southwestern United States, may have achieved a day-neutral flowering response and the annual habit as early as the first century A.D., while the same pattern in *G. barbadense* was not achieved until the late eighteenth century with the development of Sea Island cotton. The adoption of the annual habit requires a plant that grows and fruits rapidly, within a single growing season, and, moreover, one that is able to flower and fruit in the long days of the temperate-zone growing season. The latter requirement necessitates a major reorganization of the genotype, because most Gossypia, when grown under temperate-zone summer conditions, remain vegetative.

The second Old World cultigen, *G. arboreum*, occurs relatively to the east of *G. herbaceum* from India and Burma to Indonesia and China and in relatively recent times as far as Korea and Okinawa. Its origins are more obscure, but it is regarded as having diverged from *G. herbaceum* after the domestication of a common ancestor. Under this view, *G. arboreum* does not have a wild conspecific progenitor. Like *G. herbaceum*, *G. arboreum* is being displaced in cultivation to some extent by the tetraploid cultigens of the New World that have been introduced into the Middle East and the Far East.

Gossypium arboreum is more "highly bred" than *G. herbaceum*, being in many respects more specialized and better adapted to the purposes of domestication and agricultural utilization. Human selection has evidently been more successful here in recovering a wide range of types, including greatly modified plant habits; high productivity; large, easily harvested fruits; and a wide range of fiber qualities, not to mention variation in such things as leaf form and flower color.

The achievement of the annual habit in G. *arboreum* has also occurred relatively recently, possibly more recently than it occurred in G. *herbaceum*. The annual habit would clearly be required for the crop to move northward into China and Korea. It is known that this migration occurred in relatively recent times. Although cotton was known in China as an item of commerce at an earlier date, its cultivation in China first began about A.D. 900, and it did not become widely cultivated until after A.D. 1300 (Mayers, 1868). These cottons reached their northernmost extension in Korea at a still later date. It seems likely that selection for adaptation to the more northerly latitudes of China was taking place in the interval between 900 and 1300 and that the annual habit and requisite day-neutral response were achieved as a result of this selection.

The New World (tetraploid) cottons are also divided into two broad groups in the Zaitzevian conception, but the situation here is somewhat more complex. Generally, these two groups correspond to the two Linnaean species G. *hirsutum* and G. *barbadense*. There are, however, a number of other tetraploids, to be found in the wild, semiwild, or cultivated state, that do not readily fit into this simplified picture. G. *tomentosum*, which is endemic to Hawaii and evidently innocent of any history of cultivation, is generally regarded as specifically distinct from the other tetraploids. It is isolated both taxonomically and geographically. The tetraploid cotton indigenous to the Galápagos Archipelago is regarded either as only varietally distinct from G. *barbadense*, since the two interbreed freely, or as a distinct species, recognizing its morphological, ecological, and geographical distinctiveness. If it is retained within G. *barbadense* as G. *barbadense* var. *darwinii*, it must be with the understanding that it is clearly distinct in ecological terms, since it is the only unquestioned wild representative of this species. Whether it can be regarded as a wild progenitor or not is unclear.

The rare wild cottons of northeastern Brazil, here treated as G. *mustelinum*, may be included in another distinct tetraploid species, one that shows some resemblance to G. *tomentosum* of Hawaii (for example, in indumentum), in spite of their wide geographical separation. In many respects *G. mustelinum* is intermediate between *G. hirsutum* and *G. barbadense*, but it appears to deserve recognition in specific rank (cf. Pickersgill, Barrett, and de Andrade-Lima, 1975). The distinctive plant from western Mexico *G. lanceolatum* has, in most recent treatments, been submerged in *G. hirsutum*. Johnson (1975), however, has recently presented evidence to suggest that *G. lanceolatum* (as *G. hirsutum* var. *palmeri*) and *G. hirsutum* are amphidiploids of different parentage and separate evolutionary origins and thus should be recognized as distinct species, a view that is adopted here. Thus, to resolve the New World tetraploids into the Zaitzevian conception of two broadly varying species, *G. hirsutum* and *G. barbadense*, I believe it is necessary first to segregate for special consideration four segregate tetraploid species: *G. tomentosum, G. darwinii, G. mustelinum,* and *G. lanceolatum.*

The cottons referable to *G. hirsutum* show a broad range of variation, from highly specialized cultigens (the "Upland" cottons) to fully wild forms, with various intermediates. The influence of man has been very great and is in many instances readily demonstrable. Certain of the wild forms occurring as an integral part of the strand vegetation in tropical America may be taken as a point of departure. Their characteristics have been variously modified and capitalized upon by man in developing a plant better suited to his needs as a cultigen. Let us attempt to follow some of those changes.

The wild plants occur as "outpost shrubs" far out on the beach where they are continually exposed to salt spray on their leaves. This tolerance of salinity has been a significant element in the agricultural success of the Upland cottons, often enabling them to be grown in soils unsuitable for other crops, such as *Citrus* spp., that are less tolerant of soil salinity.

The wild form occurs on beaches rimming the Gulf of Mexico (for example, in Tamaulipas, Yucatán, and Florida) and to some extent in the Caribbean and extends to various localities in Polynesia. It is generally a low-growing perennial shrub, often forming large clumps. The fruits are small; the lint is sparse and fine. It is a plant that might well be gathered and used by man but that would require modification to become a satisfactory cultigen. It stands in somewhat the same relation to the cultivated representatives of *G. hirsutum* as the wild *G. herbaceum* var. *africanum* bears to the cultivated representatives of *G. herbaceum*. It is an indigenous part of the native vegetation.

A second group of representatives of *G. hirsutum* is found widely through Middle America and to some extent also in northern South

America and West Africa. These are the dooryard cottons, generally growing in association with man in yards, on trash heaps, and along roadsides or in a feral state. They do not form an established part of native vegetation; those that have run feral are generally confined to disturbed sites. These plants do not show any marked morphological differences when compared to the wild littoral cottons described before, but differ primarily in ecological terms. Yet that kind of difference is considered very significant. The dooryard cottons occur inland in positive association with man; the littoral cottons are coastal and (as Sauer has pointed out) occur in negative association with man. It is for these reasons that the latter are considered primitively wild cottons and the former are not.

This interpretation finds support in one often-overlooked morphological character. It was mentioned earlier that the littoral wild cottons have relatively sparse, fine fibers on the seeds. If with cultivated cotton one measures the amount of fiber produced (as cotton breeders routinely do) and reports this measurement as "lint percentage," which is the ratio of the weight of fibers to the weight of fibers plus seeds of a given sample, one learns that about one-third of the weight is fiber and two-thirds of the weight is seed. That is to say, lint percentage in cultivated cottons is of the order of 33 percent. Of course, the character is variable, and the range roughly extends from 25 percent to 40 percent for cultivated cottons, but the variation is continuous in nature. If, however, we extend our attention to the whole range of variation within the species, including wild, feral, dooryard, and cultivated types, we find that the breadth of variation for lint percentage is indeed extended, as one would expect, but that it is no longer continuous. We find that lint percentage is distributed bimodally. The range extends downward to around 18–20 percent and tails off rapidly below that, except that a secondary peak is found at about 8–10 percent. The cottons in this second peak, with very sparse fibers, are the wild littoral cottons. They are set apart from the dooryard, feral, and cultivated cottons by discontinuous variation in a character that is agriculturally important—the amount of fiber produced per seed. This seems to me to be strong supporting evidence for the view that these littoral cottons are the kind of cotton that early man found and later domesticated and are not the result of reestablishment of formerly domesticated cottons in the wild. That is, they are primitively wild rather than secondarily so. The feral cottons that occur inland in disturbed sites and which have lint percentages of 20 percent or more are secondarily wild and reveal their past histories in these characters.

Domestication in G. hirsutum, thus was evidently accompanied first by selection for increased quantity of fiber on the seed. Selection for increased fruit size evidently came much later, because many of the dooryard and feral cottons have small fruits of a size comparable to the fruits of the wild littoral cottons. Large-fruited types of G. hirsutum were eventually selected, but this change probably followed the development of more intensive agricultural practices.

The development of intensive agriculture and monoculture of crop plants, including cotton, has led to increased problems with pests and diseases. The crop has had to evolve in response to these challenges, and one common response has been the development of earlier and more prolific fruiting habits. In other cases the crop has had to develop some specific characteristics in response to a particular pest—for example, increased hairiness to counteract the depredations of jassids or increased earliness after the invasion of the U.S. Cotton Belt by the boll weevil in the early part of the twentieth century.

The industrial revolution saw the parallel developments of the cotton gin, which permitted larger-scale production of cotton, and the spinning jenny and associated innovations, which permitted the larger-scale consumption of cotton. These developments placed additional requirements on cotton as a crop. Premiums were placed on fiber quality, which usually can be translated into terms of fiber length and fiber strength, and on the uniformity of the product. These new requirements were translated into new types of plants as the cotton breeders hybridized, selected, and guided the evolution of the crop into new pathways. Some of these patterns are indicated in Tables 1 and 2.

Recent developments include the large-scale introduction of mechanical harvesters, requiring modifications in the growth habit of the plants, and increasing concern with the problems of pesticide residues in the biosphere, which has given rise to research aimed at developing host-plant resistance to various pests and diseases to replace or at least reduce the use of toxic pesticides.

A look to the future reveals the potential and increasing importance of cottonseed as a human foodstuff. Cotton seeds contain large amounts of high-quality edible oil and a high proportion of protein. The

Table 1. Effects of Changing Requirements on the Cotton Crop.

Selection Pressure	Response	Side Effects
First planted by man (need for reliable germination)	increased lint density larger fruits larger seeds permeable seed coats	— fewer fruits greater susceptibility to predators reduced adaptation to the wild
Requirements of hand ginning	lint-fuzz differentiation kidney-seed trait	— loss of reproductive efficiency
Industrial revolution (invention of the spinning jenny, ca. 1770)	larger, more dependable yield increased uniformity of product	susceptibility of monoculture to pests and diseases
Invention of Whitney's cotton gin (1795) and Macarthey's roller gin (1840)	increased productivity	further development of pests and diseases
Introduction of boll weevil to U.S. Cotton Belt (1900)	earlier varieties	loss of fiber quality
Introduction of synthetic fibers (1950)	increased fiber strength	loss of yield
Introduction of mechanical harvesting (1960)	return to smaller fruits new cultural practices	more fruits reduced plant stature
Development of seed protein technology	de-emphasis on fibers (?)	?

Table 2. Changing Patterns for Selected Characters through the Course of Cotton Domestication.

Character		Period of Cotton History			
	Wild	Dooryard	Post–Industrial Revolution, Pre–Cotton Gin	Post–Cotton Gin	Modern Intensive Agriculture
seed coat	impermeable	either?	either?	permeable	permeable
fiber–fuzz differentiation	poor	weak	moderate	strong	strong
kidney-seed trait	undesirable	desirable	undesirable	undesirable	undesirable
fiber density	low	high	high	high	high
fruit size	small	small→ large	large	large	large→ small
productivity	low	low	high	high	very high
growth habit	perennial	perennial	either	annual	annual
lint detachability	difficult	easy	easy	easy	easy
seed size	small	small or large	large	large	small
habitat	littoral	disturbed	cultivated	cultivated	cultivated

oil has been extracted and used as an important by-product for many years, but the technology for extracting and preparing the protein in usable form is only now being developed on a commercial scale. With increasing human population and decreasing food supplies, this use of the cotton crop may be expected to assume greater importance, the food value of the crop perhaps eventually surpassing its value for fiber. Should this change occur, the evolution of the crop may take new, as yet unpredictable directions.

An interesting sidelight into the evolution of cotton as a crop plant may be had by considering the case of kidney cotton, a specialized form of *Gossypium barbadense*, *G. barbadense* var. *braziliense*. This cotton gets its name from having the seeds of one locule (about ten) fused into a solid mass that is somewhat kidney-shaped. Under conditions of primitive dooryard agriculture, as may be found in the Amazon Basin where this plant is grown, there is a dual advantage to this condition. It is a much more efficient way to "gin" the seeds (that is, remove the fibers from the seeds), when one must gin by hand without benefit of machinery, if the seeds are fused into such a mass than if one must gin each seed individually. Also, it is of some advantage at planting time, when planting by hand, to plant each hill with a "lock" of fused seeds, with the seeds all properly oriented for optimum emergence and with the seeds (because of the greater mass) more readily placed at the proper depth. Thus, it is not difficult to imagine that this character was selected for and fixed under cultivation. It does not occur in any of the wild lint-bearing cottons.

But the kidney-seed trait was no longer advantageous when cotton agriculture moved from primitive dooryard cultivation of a few plants to large-scale field production after the Industrial Revolution. Hence, this specialized type of cotton was pushed to the side and persists today only as a remnant of a more primitive type of cotton agriculture.

And it is on this note that I wish to conclude the chapter. The myriad forms of lint-bearing cottons that exist today, of which the kidney cotton is but one specialized example, are leftover remnants of a long course of crop evolution extending from wild progenitors such as were "found," gathered, and used in some fashion by early preagricultural man to the highly evolved commercial crops of today. All the intermediate steps resulting from various (and sometimes conflicting) requirements and directions of selection and evolution have in some degree left remnants along the wayside. These, together with the surviving representatives of the wild lint-bearing cottons, present an imposing array of variation. The hand of man has been pervasive in the evolution of these plants. The variation can be understood only if we consider its historical (and prehistorical) origins—in a word, its evolution. Thus, we should not be too critical of the early botanists, such as Todaro and Watt, who did not have today's accumulated knowledge of this evolutionary development and who in consequence were overly impressed with the manifold nature of the variation. We must understand their position as we reduce their unwieldly taxonomy to more manageable proportions and, we hope, to a more accurate reflection of botanical reality (cf. Smith, 1969). In order to do this, we must first understand the influence man has had on these plants.

7

Chromosomal Patterns

The fox knows many things, but the hedgehog one big thing.
Aristolochos of Paros

CYTOTAXONOMY is a field for which much has been claimed and in which numerous successes have been scored. In many groups of plants and animals a knowledge of chromosome patterns has provided key information in clarifying species relationships and in helping us understand the probable evolutionary history of a group. But, as with other lines of evidence, cytology is not a cure-all to resolve all problems, as the more extreme of its proponents would have us believe. Cytological data provide more useful information in some groups than in others; it is with this caveat that we may begin to consider the bearing of a knowledge of the chromosomes upon the evolutionary patterns of the tribe Gossypieae.

To begin with, we find no elaborate series of different ploidy levels, such as certain other groups exhibit, that enables us to draw conclusions on relationships and to put directional arrows on the evolutionary history. Except for the few tetraploid species of *Gossypium*, there are no other examples of polyploids in the entire tribe. In the Gossypieae nearly all of the species are at a singly ploidy level.

We do not, however, find only a single chromosome number, but rather several closely related chromosome numbers $2n = 20$, 22, 24, and 26, as well as 52 for the tetraploid species of *Gossypium* (Table 3). What can we learn from these figures? First of all, by comparison with other groups of plants of comparable diversity, the tribe Gossypieae seems to be remarkably stable cytologically, especially considering the rather great diversity and presumably great age that the group has. The number $2n = 26$ seems to be numerically predominant and is characteristic of three of the four large genera that are extant: *Gossypium*, *Thespesia*, and *Hampea*. Only two genera (*Gossypioides* and *Kokia*), with a total of six species, have the reduced number of $2n = 24$. The numbers $2n = 20$ and $2n = 22$ are found only in *Cienfuegosia*. *Lebron-*

Table 3. Reported Chromosome Numbers in the Tribe Gossypieae.

Species	Chromosome Number (2n)	Species	Chromosome Number (2n)
Cienfuegosia affinis	20	G. harknessii	26
C. argentina	20	G. herbaceum	26
C. digitata	20	G. hirsutum	52
C. drummondii	20	G. klotzschianum	26
C. hearnii	22	G. lanceolatum (= palmeri)	52
C. heterophylla	20	G. lobatum	26
C. hildebrandtii	22	G. mustelinum	52
C. rosei	20	G. raimondii	26
C. somaliana	22	G. somalense	26
C. sulfurea	20	G. stocksii	26
C. tripartita	20	G. sturtianum	26
C. ulmifolia	20	G. thurberi	26
C. welshii	22	G. tomentosum	52
C. yucatanensis	20	G. trilobum	26
		G. triphyllum	26
Gossypioides brevilanatum	24		
G. kirkii	24	Hampea integerrima	26
		H. nutricia	26
Gossypium anomalum	26	H. rovirosae	26
G. arboreum	26	H. stipitata	26
G. areysianum	26	H. tomentosa	26
G. aridum	26		
G. armourianum	26	Kokia cookei	24
G. barbadense	52	K. drynarioides	24
G. bickii	26		
G. capitis-viridis	26	Thespesia danis	24
G. darwinii	52	T. garckeana	26
G. davidsonii	26	T. lampas	26, 28
G. gossypioides	26	T. populnea	24, 26, 28

NOTE: The data in this table are summarized from the literature cited by Bolkhovskikh et al., 1969, and from a few more recent reports.

necia and Cephalohibiscus are as yet unknown cytologically, but I am rash enough to postulate (see chapter 9) that the gametic chromosome number will probably be found to be $2n = 26$ for both.

One can conclude that 26 is the base number for the tribe, not only because it is the most prevalent count, but also because the morphologically less specialized genera are characterized by this number. The greatest deviation from this number ($2n = 20$) is found in the most highly specialized, herbaceous species of Cienfuegosia. Consequently, it seems sound to view the tribe as based on a chromosome number of $2n = 26$, with some representatives showing numbers that represent a reduction series below that value. The conflicting counts reported for *Thespesia* (Table 3) remain to be resolved.

It is true that aneuploids are known both above and below the original base number in various plant groups, but it is also true that reduction series below the base number are more usual. There are doubtless sound cytoevolutionary reasons for this, which I will leave to the cytologists to expound. In the present case there seems to be little room for doubt that we are dealing with an ancestral number of $2n = 26$ that has been reduced in certain instances to numbers of $2n = 24$, 22, and 20.

There are a number of species that have not yet been counted. I have already speculated that *Lebronnecia kokioides* and *Cephalohibiscus peekelii* will be found to have chromosome numbers of $2n = 26$, based on the evident similarities of these two monotypic genera to *Hampea* and *Thespesia* (cf. chapter 9). Nearly all of the species of *Gossypium* have been counted; there are no deviations from the count of $2n = 26$, nor are any expected from the few species remaining to be counted. Only five species of *Hampea* have yet been counted (all $2n = 26$), but in view of the relative morphological uniformity of the species of this genus it is likely that this count is representative of all of them. Only a few species have been counted in *Thespesia*, but again, the counts are $2n = 26$, except for a few deviant counts noted in Table 3. Deviations from this value may occur in the remainder of the genus in view of the recently reported value of $2n = 24$ for *T. danis*. Only in *Cienfuegosia*, where about half of the species have been counted, might surprises be encountered. In subgenus *Cienfuegosia* ten species have uniformly given counts of $2n = 20$; I would expect the other seven species of this subgenus to conform to this value. In subgenus *Articulata*, four species have been counted, all yielding values of $2n = 22$. I am less sure what results to expect from the remaining three species of this group. I am especially doubtful in the case of *C. gerrardii*, which shows some morphological affinity with *Thespesia* sect. *Lampas*. It is possible that *C. gerrardii* will be found to have the transitional chromosome number $2n = 24$, but perhaps $2n = 22$ is more likely. The chromosome numbers of *C. humbertiana* and *C. hitchcockii* are also unknown, and these species are both isolated taxonomically.

Some cytotaxonomists are bound to inquire, at this point, if the differing chromosome numbers of the two subgenera of *Cienfuegosia* are not a sufficient basis for their having generic rank, and the point is well taken. Perhaps they should have generic status, but I have not made this change, either in my earlier monograph of *Cienfuegosia* or in the present work, because the differing chromosome number *alone* does not seem to be a sufficiently compelling reason to make it. Whether these taxa are distinguished at the generic or the subgeneric level is, at this point, simply a matter of taxonomic opinion, and the arguments for either alternative are equally strong. Thus, for the present, at least, I prefer not to create the new genus or to interject the attendant series of new names that would be required. Nomenclatural conservatism inhibits for the present a step that might well be taken on biological grounds. I will have more on this subject later.

Thus, we find a relatively stable chromosome pattern, involving principally a reduction series from a base number of $2n = 26$. In no case is it possible to make intergeneric hybrids, so far as is known, so we have no evidence on chromosome homology between the genera. It seems likely, however, in view of the rather remarkable chromosome stability of the group, that all of the taxa are derived from a common thirteen-chromosome ancestor and that they therefore have a residual chromosome homology. It may well be possible to obtain some data on this homology by karyotype analysis and to seek distinctive patterns of chromosome size, relative arm length, and other morphological features (knobs, constrictions, and the like) that would yield significant information on evolution at the chromosomal level. With such data it would perhaps be possible to detect chromosome repatterning among the genera and to determine which chromosome(s) had been lost in those species that participated in the reduction series—and thus whether the reduction series that occurs in the tribe in monophyletic or polyphyletic. Such studies have only recently been begun (Edwards, 1977).

Gossypium has of course been more intensively studied than the other genera of the tribe, especially in its karyology and cytogenetics. Our understanding of the origin of the tetraploid species by amphidiploidy from the hybridization of an Old World diploid species and an American diploid species is based principally on cytogenetic evidence. Chromosome pairing data have been used to divide the genus into "genome groups," which in turn have provided a major class of evidence for taxonomic subdivisions of the genus into subgenera, sections, and subsections.

The detailed and elegant studies that have been made in the genus *Gossypium* not only have shown the varying degrees of homology among the different genomes on the basis of direct chromosome pairing of interspecific hybrids, but also have produced additional data from more sophisticated studies. These studies have involved the use of bridging species, where direct hybridization is impossible; the comparison of the constituent genomes of the tetraploids with those of the diploid species; the comparison in certain cases of differing chromosome end-arrangements among the different genomes resulting from translocations; and other studies involving multiple-species hybridization and the creation of various artificial ploidy levels, including triploids, hexaploids, and pentaploids. One of the principal conclusions to be derived from these studies concerns the high degree of stability, in structural terms, shown by the chromosomes of *Gossypium*. Chromosomal rearrangements (such as inversions and translocations) have occurred, but at a surprisingly low frequency considering the great age generally attributed to the group and the great amount of time that has been available for such changes to accumulate. In fact, the mystical term "cryptic structural differentiation" was coined to "explain" the puzzling absence of chromosomal rearrangements, because at that time it was believed that chromosomal evolution could not proceed without such structural changes. They were called "cryptic" because they just *had* to be there—it was just that one could not see them. Now we realize that those rearrangements that do occur are almost incidental to the process of chromosome evolution in *Gossypium* and are certainly not the dominant theme or driving force that they often are in other groups. We are again forced to acknowledge the remarkable chromosomal stability of these plants.

Studies of the other genera of the Gossypieae, comparable in depth to the cytogenetic studies of *Gossypium*, have not been made, but it seems reasonable to extrapolate that the structural stability noted for *Gossypium* is probably also characteristic of the other genera in the tribe, since the stability of chromosome numbers is characteristic of the entire tribe.

The stability of numbers for the tribe is, of course, qualified by the existence of a reduction series, producing chromosome numbers of $2n = 24$, 22, and 20 in addition to the base number of $2n = 26$. It has been suggested that the base number of 26 is itself an ancient polyploid derived from parental stocks with chromosome numbers of $2n = 12$ and

$2n = 14$. There is little evidence with which to support this speculation, although some comes from data on residual chromosome pairing
in haploid cells (Vijendra Das and Mensikai, 1968). However, the existence of a reduction series in the tribe is in itself a form of evidence in
support of the hypothesis of an ancient polyploid origin. That taxa of
the Gossypieae are able to tolerate loss of chromatin, to the extent of
one or more whole chromosomes, strongly suggests a polyploid constitution, however ancient its origin. Polyploids, by definition, have
duplication and redundancy in their germplasm and can therefore
safely withstand the loss of a chromosome, survive, and evolve. This
has evidently happened in the Gossypieae, and that it has happened
can be taken as at least suggestive that the base number of $2n = 26$ was
amphidiploid in origin, even though we have lost sight of the details of
that origin.

Nearer to us in time than the origin of the tribe itself are the
origins of the three genera that deviate from the remainder of the tribe
in their reduced chromosome numbers: *Kokia* and *Gossypioides* with
$2n = 24$ and *Cienfuegosia* with $2n = 20$ or 22. I suggested earlier that
studies of chromosome morphology might provide evidence on intergeneric chromosome homology and may possibly tell us *which*
chromosomes had been lost in these reduction series. Should we succeed in gaining such information, it will be possible to judge if these
reduction series are monophyletic or polyphyletic. For example, if the
chromosome lost in *Kokia* is different from the one lost in *Gossypioides*, we may conclude that these two genera, in spite of their
common chromosome number, originated independently. It does not
seem possible to answer this question by direct hybridization, but it
may be possible to do so by karyotype analysis. For example, J. B.
Hutchinson (1943) has shown that one chromosome of *Gossypioides
brevilanatum* is twice as long as the other chromosomes, so it may have
achieved its reduced number by chromosome fusion instead of loss.

On the basis of comparative gross morphology (as well as
phytogeography), I would opt for polyphylesis in accounting for the
reduced chromosome numbers in these three genera. It is more plausible to me to envision a derivation of these evolutionary lines from a
thirteen-chromosome progenitor as three (or perhaps four) separate
and independent reduction series. *Gossypioides* seems quite obviously
to have been derived from a *Gossypium*like ancestor, but what manner
of ancestors gave rise to *Kokia* or *Cienfuegosia*, or what their affinities
within the tribe may have been, is more difficult to see at this point.

It is possible, however, that a karyotype analysis and the subsequent determination of the identity of the individual chromosomes lost in the two taxa of *Cienfuegosia* (the American group with $2n = 20$ and the African group with $2n = 22$) could provide the crucial evidence needed to resolve the question of the rank of these taxa. Are they subgenera of *Cienfuegosia*, or are they better regarded as distinct genera? If the comparative study of chromosome morphology can show that the two groups have had an independent origin through two different reduction series involving the loss of different chromosomes, it will be convincing evidence that we are dealing with two evolutionary lineages and thus two distinct genera.

8

Evolution

Nothing in biology makes sense except in the light of evolution.
Dobzhansky, 1972

IT is possible to discuss evolutionary trends to the extent that one can successfully reconstruct probable phylogenies from comparative studies of extant taxa and to the extent that one can plausibly deduce evolutionary series of primitive and advanced traits (cf. Eyde, 1971). Such data do not provide as satisfactory a record as an actual phylogeny reconstructed from fossil evidence, but meaningful results can often be achieved even in the absence of fossil data. Some may think it idle to speculate on evolutionary rates in the absence of a fossil record for the group, but probably some limited conclusions can successfully be drawn on that subject, too.

Evolutionary Trends

OUR examination of the tribe Gossypieae reveals eight clearly defined genera. Alliances among certain of these genera are fairly clear, while other genera are more isolated. Throughout the group numerous trends, combining and interacting in various ways, are discernible. These trends are perhaps best discussed individually.

Growth habit. There is a trend toward reduction in growth habit from the arborescent forms that are primitive and typical of several of the genera to the shrubby and herbaceous forms that are derived from the arborescent habit in several lineages. Shrubby growths occur in a few species of *Thespesia* (*T. lampas* and *T. thespesioides*), both species of *Gossypioides*, and most species of *Gossypium*. Certain species of *Gossypium* are perhaps better described as subshrubs (for example, *G. populifolium* and *G. stocksii*), as are several of the species of *Cienfuegosia* (such as *C. hildebrandtii* and *C. gerrardii*). Only in the genus *Cienfuegosia* are fully herbaceous types encountered, always growing from a perennial rootstock. The extreme examples of this trend are found in the prostrate herbaceous species of *Cienfuegosia* section *Friesia*.

Ecological adaptation. The angiosperms are thought to have originated in a tropical montane habitat and to have spread from there to lower elevations and higher latitudes. Relatively primitive angiosperms are still concentrated in the montane tropics (cf. Takhtajan, 1969). The Malvaceae are basically a tropical and subtropical family, but, with the exception of a few genera like *Phymosia* and *Kearnemalvastrum* and the highly specialized Andean genera *Nototriche* and *Acaulimalva*, they occur at relatively lower elevations.

The tribe Gossypieae is typical of the Malvaceae generally in occurring principally in the tropics and subtropics at relatively low elevations. This may be said essentially without qualification for most of the species in the tribe. It is with the few exceptions that we can discern some evolutionary trends.

In *Gossypium* certain species have shown a limited ability to expand out of the tropics (as is discussed more fully in chapter 5). The Australian *G. sturtianum* occurs above the thirty-third parallel and has an ability, unusual for the genus, to withstand temperatures a few degrees below freezing when in full leaf. However, this cold tolerance is relatively limited, and in fact the species has extended its geographic distribution and its temperature tolerance by only a limited amount. More impressive is the evolutionary achievement of *Gossypium thurberi*, which has coupled cold tolerance with a winter dormancy pattern that enables it to withstand a fully temperate climate in the mountains of Arizona. It is interesting to note that the nearest relative of *G. thurberi*, *G. trilobum*, although fully tropical and quite susceptible to freezing temperature, nevertheless occurs at higher elevations (up to 2,600 m) than any other species of *Gossypium*.

The only other members of the Gossypieae to have penetrated the temperate zones are the several cultivated species of *Gossypium* that have been selected by man for this purpose. They might be called "artificial annuals" because man has forced these naturally perennial species into an early-flowering "annual" habit in order to serve his agricultural needs better. Needless to say, these species function as temperate-zone annuals only under cultivation and as occasional escapes. Nowhere are they established in the wild except as tropical perennials.

A few species of *Hampea* (for example, *H. longipes* and *H. thespesioides*) occur at moderately high elevations in the tropics, but it is impossible to ascertain with present evidence whether these are to be interpreted as relicts of an ancestral montane type or as secondarily specialized montane derivatives. That is to say, it is impossible to decide in this case which way the evolutionary arrows should point.

Now, if we leave aside temperature and elevation and consider moisture, we see a clear trend from mesophytically adapted plants to xerophytically adapted plants. The genera *Cephalohibiscus*, *Thespesia*, and *Hampea* are the most mesophytic of the tribe. In *Thespesia* a few species are adapted to dry savannas (for example, *T. thespesioides* and *T. danis*), and two species, *T. populnea* and *T. populneoides*, have become specialized by their adaptation to a halophytic environment, which in physiological terms is like a xerophytic environment in many respects. The greater part of *Thespesia*, however, is at home in tropical rain forests, especially in New Guinea and the West Indies.

Many of the species of *Hampea* also are at home in the tropical rain forests, but certain of them (for example, *H. tomentosa* and *H. mexicana*) occur in regions that are subject to a rainy-season–dry-season cycle that exposes them to severe drouth during a major part of the year. These species, however, show no obvious adaptation to this environmental factor either in their structure or in modifications of their growth pattern or life cycle.

The oceanic genera *Kokia* and *Lebronnecia* and the African *Gossypioides* show a somewhat greater degree of adaptation to a drier habitat, but the number of species is too few to discern trends.

Cienfuegosia is adapted to a variety of habitats of varying degrees of xerophytism. Several species, including the more primitive representatives of the genus (such as *C. affinis* and *C. gerrardii*), are savanna plants. Other adaptations are consequently regarded as derived from the savanna adaptation. Two species (*C. heteroclada* and *C. integrifolia*) seem to be especially adapted to regularly burned-over savannas by flowering directly from the perennial rootstock and fruiting rapidly in a narrow seasonal pattern. Other species (primarily in section *Friesia*) are adapted to saline conditions, often occurring in areas that are both saline and arid, frequently on sandy soils. The species of section *Paraguayana*, on the other hand, commonly are found on heavy soils in situations that might best be characterized as "wet meadows."

In section *Cienfuegosia* we find a pattern that, although difficult to pin down, is suggestive of saltwater dispersal. The distribution of *C. yucatanensis* is clearly achieved by saltwater dissemination; the species occurs in coastal areas along the margins of the Gulf of Mexico—in

Yucatán, Florida, Cuba, and the Bahamas (Fig. 52). The remaining species of the section are widely dispersed from Mexico to Paraguay to Africa. Although all are basically continental in their occurrence, all (except *C. subternata*) occur to some extent in more or less coastal areas. Thus, their wide disjunction, together with their close relationship genetically (all interbreed easily, so far as they have been tested), suggests an extreme means to achieve this dissemination. Their general proximity to coastlines (if not their actual occupancy of the littoral zone) and the clear ability of *C. yucatanensis* and *C. heterophylla* (which also island-hops) to be dispersed on ocean currents make it highly likely that saltwater dispersal has been a major factor in the spread of the several species of section *Cienfuegosia*. Thus, it seems that these plants, in evolving a more xerophytic adaptation, have been able at least to pass through, if not occupy, the littoral zone.

Finally, the most extreme climatic adaptations in the genus *Cienfuegosia* are found in those species that occur in desert areas. These include the Arabian species of section *Garckea* and the Peruvian-Ecuadorian *C. tripartita*. The Arabian and the coastal Peruvian deserts are among the severest in the world. *C. digitata* also may be found in extremely arid areas in Mauritania, Senegal, and Mali. The Arabian species show definite xeromorphic adaptations to their environment, such as sclerophyllous leaves, but the same is not true of *C. tripartita*. The distribution of *C. tripartita*, however, is not only in the extremely dry coastal areas of northern Peru and southern Ecuador, but also somewhat inland in northern Peru along the Río Marañon under much more moist conditions. One is tempted to conclude, in view of the clear lack of xeromorphic adaptations of *C. tripartita*, that in evolutionary terms it is in the process of extending out to the west from its more mesophytic environment along the Río Marañon. The perennial rootstock is evidently sufficiently capable of withstanding prolonged drouth that the species has become established in this arid zone on the basis of the growth it is able to achieve on those relatively rare occasions when moisture becomes available. Very likely this "push to the coast" (and to the island of Puná) provided the opportunity for long-distance saltwater dispersal that was responsible for the establishment and differentiation of the closely related *C. rosei* in coastal Oaxaca.

Comparably complex trends of adaptation may be seen in the genus *Gossypium*. The majority of the wild diploid species are more or less xerophytically adapted. The least so are the phylogenetically isolated *G.*

longicalyx from East Africa and the several little-known species from the north coast of Australia (for example, *G. cunninghamii* and *G. pulchellum*). The latter are subject to large amounts of rainfall, but the rainy season in that area alternates with a dry season, so it is difficult to characterize these species as mesophytically adapted. Perhaps they are best described as savanna plants. *G. longicalyx* occurs in an area that is only somewhat more mesophytic than the areas in which most diploid species of *Gossypium* occur.

The wild cottons typically occur in desert areas. In Africa these plants may be found in Angola and Namibia (South-West Africa) and in a zone from Mali to Somalia. They extend to the east onto the Arabian peninsula and (in the case of *G. stocksii*) as far as Sind, Pakistan; to the west *G. capitis-viridis* occurs in the Cape Verde Islands.

In Australia, with the exception of the species previously referred to on the nothern coast, the remaining species are characteristic of the interior deserts. The only exception to this distribution is *Gossypium sturtianum* var. *nandewarense*, which occurs in a relatively more mesophytic habitat than var. *sturtianum*. The occurrence of var. *nandewarense* in a few mesophytic pockets is interpreted as a relictual distribution, and var. *sturtianum* is regarded as having evolved the more xerophytic phenotype in response to the increasing aridity of the Australian landscape that has occurred since the late Quaternary.

In America there is somewhat greater diversity of habitat as well as greater phenetic diversity. The most extreme xerophytes are the three shrubby species of subsection *Caducibracteolata*—*G. harknessii*, *G. armourianum*, and *G. turneri*—which have evolved various xeromorphic adaptations, including a double palisade in the leaves, and which in certain areas are dominant members of the vegetation. In a much different way the three arborescent species of subsection *Erioxylum* are also adapted to aridity. The widespread *G. aridum* and its more narrowly distributed relatives *G. lobatum* and *G. laxum* do not show any notable xeromorphic adaptations but have adjusted their life cycles to a rather extreme degree. These species occur in areas subjected to the wet-season–dry-season cycle. They are in leaf only during those months when moisture is plentiful and are completely leafless for about half the year. At about the time the leaves are lost, floral initiation begins, and flowering and fruiting ensue during the height of the dry season while the trees are otherwise dormant. *G. aridum* is both widely distributed and locally abundant; *G. lobatum* and *G. laxum*, in spite of their restricted overall distributions, are locally abundant (Fryxell and Koch, 1978); G. *laxum* is a subdominant in the vegetation of Zopilote Cañon, where it is endemic.

Other American species of *Gossypium* show less obvious adaptations to aridity, but all seem to be in some degree so adapted. Many of them (for example, G. *thurberi*, G. *davidsonii*, and G. *raimondii*), while growing in desert areas, are frequently found in or near the watercourses. However, they may also be found (as are G. *thurberi* and G. *gossypioides*) on steep, rocky slopes far from any watercourses. Evidently their principal adaptation is an aggressive and deep-penetrating root system that enables them to exist under these conditions rather than any xeromorphic adaptations for reducing water loss or any alterations of the life cycle to circumvent the aridity of the habitat. Perhaps G. *trilobum* is the least xerophytic species of *Gossypium* in its distribution, and, as has already been noted, it also occurs at the highest elevation for the genus.

Floral Biology

THE trends in floral morphology are predominantly reductional, coupled with an increasing complexity of the inflorescence. The primitive flower in the Gossypieae is large, showy, and solitary. Paradoxically, some of the largest and showiest flowers in the tribe (*T. beatensis* and *T. grandiflora*) are themselves specializations of another sort. But I am getting ahead of the story.

We may begin with the flowers of *T. populnea* or *T. lampas*, both of which are frequently grown in various parts of the tropics as ornamentals and are consequently fairly well known. These flowers are large (with petals more than 5 cm long), funnelform, and solitary in the axils. The petals are bright yellow, usually with a prominent dark maroon spot on the claw. Nectar is produced copiously by a nectary at the base of the calyx. The flowers are not odoriferous. This syndrome of characters is readily recognized as one that is attractive to certain insect visitors which serve as pollen vectors. Pollen grains in the Gossypieae are generally large and more or less sticky, adhering to one another in clumps. Such pollen is quite incapable of wind transport. Pollen vectors are therefore necessary to effect cross-pollination.

This basic pattern has been modified in a variety of ways, most of which presumably are examples of coevolution of the flowering plant and its insect (or other) pollen vector. The importance of this kind of evolutionary interrelationship has received renewed and well-deserved emphasis in recent years. It is unfortunate that observational data on pollinators of the Gossypieae are relatively meager at the present time and that most of our inferences must be drawn from floral structure and presumed adaptational patterns rather than from actual observational data. Nevertheless, several trends seem clear.

The inflorescence as a whole has changed in two different ways—and remained unchanged as isolated axillary flowers in certain groups. In the most specialized genus, *Cienfuegosia*, the inflorescence, paradoxically, has evolved the least; in section *Cienfuegosia* the flowers are solitary in the axils, and in section *Articulata* they may be either solitary or in weakly developed sympodial inflorescenses.

Such sympodial structure is well developed in *Gossypioides* and in many species of *Gossypium* and is weakly developed in *Lebronnecia*. In *Gossypium*, solitary, axillary flowers characterize two of the three species of the primitive subsection *Erioxylum*, the four species of section *Grandicalyx*, and the three species of subsection *Caducibracteolata*. Sympodial growth is only weakly developed in the four species of subsection *Pseudopambak* but is more or less well developed in the remaining species of *Gossypium*, including the cultigens. It must be noted, of course, that under suboptimal growing conditions the development of the inflorescence is restricted, and a species that potentially is capable of producing sympodial inflorescences may have these reduced to the uniflorate condition — that is, to solitary axillary flowers.

There seem to be two opposed trends here in *Gossypium*: a development from a primitive condition (exemplified by the arborescent species of subsection *Erioxylum*) toward sympodial inflorescences, and a reduction from sympodial inflorescences to the simpler condition in groups adapted to extreme aridity, such as subsection *Caducibracteolata* (principally from Baja California) and subsection *Pseudopambak* (from southern Arabia and adjacent areas). The uniflorate inflorescences of section *Grandicalyx* are interpreted as primitively so, though this is quite conjectural. In subsection *Erioxylum*, when there is a deviation from the solitary, axillary flower (such as *G. lobatum*) it is toward a fasciculate, not sympodial, grouping of flowers.

This fasciculate grouping of flowers is the second trend and is found principally in the genus *Hampea*, although, as just noted, it is expressed to a limited extent in at least one species of *Gossypium*. It is instructive to note that in the more primitive representatives of *Hampea* (that is, section *Standleya*) the fascicles are few-flowered, commonly uniflorate, whereas in the remainder of the genus (the more derived species) the fascicular pattern is more highly developed, with many flowers in each fascicle. I regard this difference as strong evidence that this pattern was derived from the solitary, axillary flower. The significance of the branched peduncle that occurs in *H. micrantha* remains problematical.

It is not clear at the present time what relationship might be suggested between these two divergent trends of inflorescence structure—whether the fasciculate pattern was developed by reduction from the sympodial, whether the reverse process occurred, or whether each pattern is an independent development from the primitive solitary flower. I tend to favor the third view at this point, but developmental-morphological studies of certain of these species could shed considerable light on the question and possibly revise this interpretation.

The primitive pattern of solitary axillary flowers characterizes the genera *Thespesia* and *Kokia* and, as already noted, many species of *Cienfuegosia* and some of *Gossypium*. I would describe the inflorescences of the monotypic genus *Cephalohibiscus* as having a modified sympodial structure; the flowers occur in pairs on long axillary peduncles.

The structure of the individual flower also shows discernible trends, most of them reductional. The type of flower considered primitive for the tribe was described earlier using the examples of *Thespesia populnea* and *T. lampas*. This type of flower characterizes most species of *Thespesia*, those of *Kokia,* many of those of *Gossypium*, and (in my view—perhaps surprisingly) most of those of *Cienfuegosia*.

Consider the flower typical of *Cienfuegosia*. It shares all of the characters regarded as primitive for the tribe Gossypieae except that of size. It usually is solitary and showy, typically has yellow petals, and usually has a dark throat. In most cases it is clearly adapted to insect pollination, a fact that is made especially evident where the style is notably elongated to present the stigma at a sufficient distance from the androecium to effectively prohibit pollination except through the agency of an insect pollen vector. Thus, the typical *Cienfuegosia* flower shares most of the essential features of the typical *Thespesia* flower except size. Petals of *Cienfuegosia* (except *C. hitchcockii*) rarely exceed 4 cm in length and are more commonly in the range of 1.5–3 cm long. Yet, if one considers the greatly reduced plant size that characterizes *Cienfuegosia* in comparison with the other genera of the Gossypieae, one sees that although absolute flower size is indeed reduced in *Cien-fuegosia*, the size of the flower *relative to the size of the plant as a whole* is not reduced. It is for this reason that I consider the floral morphology typical of *Cienfuegosia* to be primitive, in spite of the relatively specialized plant habit that it has evolved.

What evolutionary trends in floral biology can we discern within *Cienfuegosia*? We are first struck by the dichotomy in types of stigmas: decurrent vs. capitate (Fig. 50). The tribe Gossypieae is characterized by having decurrent stigmas, which give a clavate form to the combined style and stigma. On this basis we may accept the decurrent stigma as primitive for (and characteristic of) the tribe and the capitate stigma, which is found only in a part of the genus *Cienfuegosia*, as derived.

In *Cienfuegosia*, subgenus *Articulata* has decurrent stigmas, whereas subgenus *Cienfuegosia* includes both types of stigmas, mostly decurrent in section *Cienfuegosia* and capitate in sections *Robusta*, *Paraguayana*, and *Friesia*. Here I believe I am justified in postulating a reversal of evolutionary trend. The two subgenera are sharply distinct and clearly natural taxa; indeed, the argument may be made that they deserve generic rank (see chapter 7). It seems plausible to assume that the dichotomy in type of stigma originated with the divergence of subgenus *Cienfuegosia* from the remainder of the Gossypieae, but that in evolutionary terms the achievement of the capitate stigma was not stable and reversion to the decurrent stigma occurred in one evolutionary line within the subgenus but not in the other three. In support of this view I note that within section *Cienfuegosia*, where decurrent stigmas are characteristic and where the component species are quite closely allied (indeed potentially interbreeding), subcapitate stigmas occur in *C. subternata* and occasionally also in *C. tripartita*. Since this section, together with section *Robusta*, has the relatively more primitive species and section *Cienfuegosia* the relatively more derived, it seems evident that the evolutionary trend here is from the capitate to the decurrent stigma and that this pattern constitutes a reversal of trend. Presumably, the genetic basis for this difference in stigma type is relatively simple and therefore easily influenced in evolutionary direction by selection. The genetic control of this character should be testable, since the several species of section *Cienfuegosia* are all interfertile insofar as they have been tested; unfortunately, *C. subternata* with subcapitate stigmas is not yet available in experimental culture.

There appears to have been a strong selective pressure in *Cien-*

fuegosia for outbreeding, since many species, whether with capitate or decurrent stigmas, have evolved notably elongated styles. Yet in most cases the outbreeding remains facultative. The decurrent stigma itself is a device for permitting at least limited self-pollination, should cross-pollination fail. Only in the species with capitate stigmas does there seem to be an evolutionary attempt at obligate outbreeding. And even there, certain species (for example, *C. sulfurea* and *C. argentina*) have evolved a divided style, the branches of which recurve (much as in *Hibiscus*) in a manner that enhances the possibility of self-pollination, should cross-pollination fail. As far as is known, all species are self-fertile. The general trend, therefore, seems to be toward facultative outbreeding.

The exceptions to this general trend are, of course, instructive. In *C. heterophylla* the style is sufficiently shortened that the decurrent stigma is essentially surrounded by the numerous anthers. Moreover, the petals do not fully open at anthesis. Under these circumstances self-pollination occurs easily, and cross-pollination is probably rare. In *C. yucatanensis* a marked metamorphosis of floral morphology has occurred, the significance of which is not yet clear. The flowers are the smallest in the genus, and the corollas open to a fully rotate form. It is the only species to exhibit this character. Moreover, the petal spot is completely suppressed, although all other species of section *Cienfuegosia*, even the autogamous *C. heterophylla*, have a very prominent petal spot. The anther number is notably reduced. Although *C. yucatanensis* is quite capable of setting abundant seed by self-pollination (as witnessed during my experience in growing the plant in greenhouse culture in isolation from any pollinators), the remarkable shift in floral morphology to this syndrome of characters does not seem to be an adaptation for autogamy. Rather, an adaptation to a locally available pollinator is a more plausible explanation. Unfortunately, field observations on pollinators of *C. yucatanensis* are at present lacking.

The most extreme deviation from the general pattern of floral biology in *Cienfuegosia* is found in the dioecious *C. heteroclada* (Fig. 8), which of course is obligately outbred. Unfortunately, we have only limited knowledge of even the floral morphology in this species, much less of the genetic control of dioecism, of the population dynamics of the two sex forms, or of any interactions with pollinators. Such interesting knowledge awaits further study of this little-known West African species, but the simple fact of the occurrence of dioecism is both exceptional and noteworthy.

But let us now consider other trends of specialization from the basic pattern of the tribe. There are several themes of variation on this basic pattern. Dioecism is one such theme. It is fully developed in *Hampea*, in which all but three of the species (section *Standleya*) are fully dioecious. It occurs in *Cienfuegosia heteroclada* in what must be regarded as an isolated occurrence of this phenomenon. That it is found in this isolated fashion suggests, however, that selection can establish this condition without too much difficulty. The occurrence of male sterility (or gynodioecism, or anther abortion) in natural populations of *Cienfuegosia rosei* (and possibly also in *C. affinis*) may be an example of a first step in the breakdown of the hermaphroditic flower that could readily lead to the dioecious condition under the appropriate selection pressures. Limited data (Fryxell, unpublished) indicate that the male sterility in *C. rosei* is transmitted maternally, presumably under cytoplasmic control. What selection pressures maintain this sexual polymorphism are not understood.

In *Hampea* spp., staminate and pistillate individuals occur in roughly equal proportions in the field. In *Cienfuegosia heteroclada*, pistillate individuals make up only 10 percent of natural populations. In *C. rosei*, male-sterile individuals make up approximately one-third of natural populations. Clearly there are differences here, either in the genetic control of sex expression or in the selective processes at work on individuals of different flower types, about which we know very little at present.

It is in *Hampea* that we find the greatest restructuring of the floral morphology. There, in addition to dioecism, we find a reduced flower size; a reduced fasciculate, primary inflorescence but an extended secondary one; reflexed, somewhat fleshy petals; a loss of petal spot and an absence of bright coloration of the petals; and the presence of a floral odor. Many of these traits are related to questions of pollination and the attraction of pollen vectors. The overall syndrome of floral biology of *Hampea* suggests adaptation to crepuscular pollinators such as moths (cf. Baker, 1961), in contrast to the more brightly colored (yellow, red) flowers of most Gossypieae, which have darkly pigmented spots on the claws of the petals (the "honey guides") in a pattern associated with diurnal pollinators such as bees. Extensive data on the activities of pollinators in *Hampea*, however, are not yet available.

This differing pattern of floral biology, which sets *Hampea* off from the other genera of Gossypieae (with the possible exception of *Lebronnecia* and *Cephalohibiscus*), may well relate to its existence in a different ecological setting. The wet forest where *Hampea* is at home presents different pollination problems and opportunities, and presumably different potential pollinators, than do the desert and the thorn scrub where *Gossypium* occurs or the savanna where *Cienfuegosia* is found. A comparison with the pollination patterns of the forest-dwelling *Cephalohibiscus* and the *Thespesias* endemic to the forests of New Guinea might well prove to be enlightening.

It is interesting that the apparent adaptation of the small-flowered *Hampea* to crepuscular pollinators is apparently paralleled by the adaptations of the large-flowered *Thespesia beatensis*. This species, which is narrowly endemic to a very small Caribbean island, Isla Beata, has large, creamy white flowers completely innocent of any spot at the base of the petals. A single specimen of this tree, introduced by the Fairchild-Dorsett expedition of 1932 (cf. Howard, 1949), has been maintained at the USDA Plant Introduction Garden in Florida for many years. I procured a portion of this plant (rooted as a marcot; members of the Gossypieae root as cuttings only with difficulty) and grew it in the greenhouse for several years before losing it to a greenhouse accident. During that time it flowered freely. I made countless hand pollinations, but no fruits or seeds were produced. (I am unaware that any fruits have been produced on the Florida tree, either.) This failure to fruit suggests that there may exist some form of self-incompatibility in *Thespesia*, since I have had a similar experience with isolated individuals of *T. garckeana* and *T. populneoides*. Other species of *Thespesia* (*T. populnea, T. lampas,* and *T. thespesioides*) and all of the species of *Gossypium* and *Cienfuegosia* that have been grown experimentally are not known to exhibit any self-incompatibility. But the possible presence of such in some species of *Thespesia* and the presence of dioecism in *Hampea* (both those conditions requiring pollen vectors) point emphatically to the adaptational considerations that must be brought into discussions of floral biology. The parallel of *T. beatensis* and *Hampea* spp. in their apparent adaptation to crepuscular pollinators is therefore difficult to disregard.

Gossypium tomentosum provides yet another instructive example. Although individuals of this species are fully self-compatible, they fail to set seed in isolation from pollinators because of a very long style that elevates the stigmatic surface too far above the androecium for spontaneous self-pollination to occur. In the greenhouse, hand pollination is required; in the field, pollen vectors are required and evidently are effective, but field observations are lacking. The evidence suggests that for this species, too, crepuscular pollinators (presumably moths) are involved in the adaptational pattern.

In addition to the very long style, the pattern includes the complete absence of the dark petal spot, a distinctive glossy yellow petal color (notably visible in subdued light), the unusual ability of the flowers to remain open through the night (whereas typical *Gossypium* flowers open early in the day and wither at the end of the same day), and a weakly developed floral odor. The last characteristic is a bit controversial in that some people (including myself) claim under favorable circumstances to be able to detect a faint floral odor in this species, while others emphatically declare that it is odorless like all the rest of its congeners. Clearly, a more sensitive instrument than the human nose is needed to resolve this question. The odor, if present, is certainly faint, but it is also unique in this one species of *Gossypium*, and its presence is consistent with a pattern of adaptation to moth pollination.

The only other species lacking a petal spot are *Gossypium longicalyx*, *Cienfuegosia yucatanensis*, *Cephalohibiscus peekelii*, and *Thespesia cubensis*. The first two species have relatively reduced flowers and other characters suggestive of a loss of dependence on pollinators and a change toward a pattern of self-pollination. The last two species are large trees with pale-colored flowers and exserted androecia. Nothing is known of their pollinators or floral biology, but the comments made earlier concerning *Hampea* (p. 195) and its adaptation to a different ecological setting may apply equally here.

Yet another variation on the basic floral pattern is the occurrence of facultative cleistogamy in certain species of *Gossypium*, specifically *G. australe*, *G. bickii*, and *G. somalense*, and in the species of *Cienfuegosia* sect. *Friesia* and sect. *Paraguayana*. It apparently does not occur otherwise in the tribe, although it is known from many other plant groups (cf. Uphof, 1938). Certain environmental conditions, not entirely understood, favor the development of these greatly reduced cleistogamous flowers, which are totally self-pollinated in the bud by a very limited number of anthers but which produce an essentially normal fruit with a full complement of seeds. When conditions become more "favorable" or "normal," the plant produces "normal" chasmogamous flowers that open fully and attract pollinators in the usual way. The two types of flowers are vastly different in size and morphology and show no intergradations. Both types may be produced simultaneously on the same plant. In *G. nelsonii* this trend has been pushed so far that cleistogamous flowers are the rule and chasmogamous flowers extremely rare.

One major trend is found in the genus *Gossypium* that is associated with the differences noted earlier in the development of the inflorescence in this genus. There is an inverse correlation of flower size with degree of development of the inflorescence. Thus, the small-flowered *G. thurberi* and *G. trilobum* have many-flowered sympodial inflorescences, while the large-flowered *G. laxum* and *G. armourianum* have solitary, axillary flowers. This pattern is paralleled in *Hampea*, in which reduced flower size is associated with a more fully developed secondary inflorescence. There are, of course, exceptions to this pattern, and each species must be considered as an individual case, but there is nevertheless merit to the generalization. The most extreme cases of floral reduction in *Gossypium*, however, are found in the African species, of which *G. somalense* may be taken as an example. In this species the corolla is small and nearly concealed by the involucral bracts; it is narrowly funnelform and pendent. In spite of these reductions, it retains the dark throat and elongated style that mark it as a species that "uses" insect pollen vectors to achieve facultative outbreeding. The reduction in size and conspicuousness of the flower presumably relates to the specific pollinators that are available in its part of the world. This view is supported by the fact that in other groups of plants from the same area (Arabia-Somalia), similar floral phenotypes are encountered: a relatively small, narrowly funnelform corolla that always has a dark throat, and often a rose or purple overall coloration. Examples that are known to me include species of *Cienfuegosia* and of *Senra*, but doubtless many others could be added. Such a phenomenon, the parallel evolution of similar floral phenotypes in essentially unrelated plants, speaks of response to similar selection pressures. It is more reasonable to conclude that these pressures concern available pollinators that are common to these plants rather than climatic or other geographic factors that they share. It is unfortunate that observational data on pollinator activity on these plants is lacking at the present time.

Indeed, it is fair to conclude that with two exceptions the pattern of dependence on diurnal insect pollen vectors is constant within the genus *Gossypium*. The two exceptions, *G. longicalyx* and *G. tomentosum*, although quite unrelated and from opposite sides of the earth, share a similar floral phenotype. They have bright yellow corollas wholly devoid of a central petal spot. Presumably, they, too, are dependent upon pollen vectors—certainly so in the case of *G. tomentosum* with its greatly elongated style—but upon a different class of pollen vectors, possibly lepidopterans instead of hymenopterans. Again we encounter the frustration of a lack of field observations of pollinator activity on these plants.

This leads us back to the large, showy flowers of *Thespesia grandiflora* and *T. beatensis*, both of which have flowers on the order of 15 cm in diameter. The former species is cultivated for its flowers, and the latter deserves to be. Both deviate from the "basic" pattern for the tribe in other floral characters besides the very large size of the corollas. Both have lost the basal petal spot, and both have lost the yellow pigmentation characteristic of a majority of the species in the tribe. In *T. beatensis* the corolla is fully rotate and essentially white or cream-colored, a pattern commonly associated with pollination by crepuscular, hovering moths, but no observations on pollinators in its native habitat (Isla Beata) have been reported. Its elongated androecium would deposit pollen on the body of such a moth, and its apparent self-sterility argues for dependence on a pollen vector. Indeed, this self-sterility may be the basic reason for the relict status of this species.

The corollas of *T. grandiflora* are reddish-colored, a character found also in *Kokia* spp. Presumably this character, too, is an adaptation to specific pollinators, but again, observational field data are lacking. The kinds of pollen vectors that these large red flowers are seeking to attract are not known.

We may conclude this discussion of floral biology in the Gossypieae by noting two species that are exceptional in having the androecium and the style and stigma prominently exserted from the corolla. In *Thespesia cubensis* (Fig. 36) this exsertion is achieved as much by reduction in the size of the petals as it is by an elongation of the androecium, but it is the *relative* size of these parts that is important in terms of the pattern presented to potential pollen vectors. In *Cephalohibiscus peekelii* (Fig. 6) the petals are curiously contorted and relatively reduced in size so that the androecium and the stigma above it are prominently presented. The pattern in these two species is superficially similar to that found in such plants as *Malvaviscus* spp., *Periptera* spp., and certain species of *Hibiscus*, such as *H. poeppigii* (Sprengel) Garcke. In these plants, however, the corolla is tightly tubular and bright red and the exsertion is much greater; the adaptation is evidently to pollination by hummingbirds. The floral phenotype is actually more similar to the Mexican *Anotea flavida* (DC.) Ulbrich and the Brazilian *Pavonia macrostyla* Gürke, because they have pale yellow corollas, although their androecia are much further exserted than in *Thespesia cubensis* or *Cephalohibiscus peekelii*. It is not known what the pollinators of *T. cubensis* or *C. peekelii* are, but the floral character syndrome is different from that of the plants pollinated by hummingbirds, so presumably the pollinators are different. The exsertion of the androecium and stigma is the only common character, and its significance is not yet clear.

In sum, the Gossypieae show a wide range of variability in their floral adaptations. Their flowers range from very large to relatively small. Dioecy, gynodioecy, self-sterility, and facultative cleistogamy are all known, although the basic breeding system evidently involves a facultative inbreeding-outbreeding pattern. The number of variations that have been imposed on this basic pattern in the evolution of the tribe is striking.

Evolutionary Rates

It was stated earlier that any discussion of evolutionary rates could only be speculative in the absence of a fossil record, because a discussion of *rates* implies some more or less precise knowledge of the time scale, which can only be supplied by a knowledge of the fossil history of the group. But is this strictly true? Even in the absence of fossils, it is possible to infer *some* things about the time scale involved.

In the present case, the tribe Gossypieae, we can tie the evolution of the group to the time scale at several points and thereby make some inferences concerning evolutionary rates. We may begin by focusing our attention on the genus *Gossypium*, for which the most clear-cut examples and the fullest information are available.

Gossypium is notable in having a rather large number of sibling species pairs or species groups. These pairs of taxa usually occupy adjacent or adjoining areas and are only sufficiently differentiated to be closely similar species, opinion being divided in some instances on whether the pairs are specifically distinct or only varietally so. I have argued in an earlier work (Fryxell, 1965b) that these sibling species pairs, of which I enumerated eight, diverged in response to the climatic change that marked the Pleistocene-Recent transition at about eleven thousand years ago. That change was to a worldwide climate that was at once warmer and drier. In the arid zones where the diploid *Gossypium* species are found, increased aridity presumably resulted in range restrictions of many species as conditions became less favorable. Previously widespread species broke up into more localized populations and, in isolation, went in different evolutionary directions. If this argument is sound, as I believe it is, we may conclude that perennial plants such as these species of *Gossypium* can diverge (in isolation) from a common ancestor by an amount such as taxonomists generally recognize as distinct species — but little more than that amount of divergence — in a period of about eleven thousand years. This is one estimate of evolutionary rate.

A second estimate can be obtained from a consideration of the origin of tetraploidy in *Gossypium*. As I have indicated elsewhere (Fryxell, 1965*b*), the origin of amphidiploidy saw the invasion not only of a new ploidy level but also of a new ecological habitat as the newly formed tetraploids occupied and dispersed themselves through the littoral zone. Since an amphidiploid, by definition, is younger than the two lineages that it combines, the origin of the amphidiploids and their invasion of the littoral habitat must be a relatively more recent event. It has been suggested that this event occurred sometime during the Pleistocene, because that was a period when the alternating advance and retreat of glaciation gave rise to a raising and lowering of sea level, which in turn produced wandering coastlines as higher sea levels moved the coastlines inland during interglacial periods and lower sea levels moved coastlines outward during glacial advances. It has been estimated that sea level fluctuated as much as 150 meters during the Pleistocene. Such mobile shorelines not only might have provided the means for bringing together (by saltwater dispersal) the Old World and New World parents of the amphidiploid in a way not yet understood, but also would have provided an open (disturbed) habitat in which the newly formed amphidiploid could become established and, once it was established, a means for it to disperse itself to littoral habitats elsewhere than its (unknown) place of origin. Such plausible reasoning suggests, then, that the amphidiploids originated sometime during the Pleistocene—that is, sometime during the last million years. This is not a very precise estimate, but it *is* an estimate of the time period during which the tetraploids Gossypia diverged to their present-day diversity, encompassing several species (six, by my interpretation).

Yet another estimate of evolutionary rates may be obtained from a consideration of the genus *Gossypium* as a whole. As presented in chapter 2 (and elsewhere), the genus is divided into three diploid subgenera that occupy three separate continental areas. It is not unreasonable to relate the divergence of these subgenera to the physical separation of the continents, that is, the breakup of the Gondwana landmass. Of course this breakup was a gradual process, so it is impossible to state precisely when the ancestral plant populations were in fact separated, but a consensus would probably agree that these subgenera are approximately of Cretaceous age—that is, approximately sixty-five million years old. Thus, we may conclude that this period of time allows the divergence of perennial shrubs to a level of almost generic distinctness, if we recognize that what a modern consensus recognizes as subgenera have in the past been treated by taxonomists as several genera.

So in *Gossypium* we have three different estimates of evolutionary rates even though there is no fossil record. Against this background, what other evidence can we find in the other genera of the Gossypieae? The six species of *Cienfuegosia* sect. *Cienfuegosia* appear to support the interpretation presented for the tetraploid species of *Gossypium*. In this case also we have a lineage that has diversified into several distinct species, but these species have not diverged to such an extent that they have lost the ability to produce fertile hybrids readily when brought together in artificial culture. Moreover, we see strong phytogeographic evidence that these species have been dispersed with the help of saltwater flotation of seeds. This interpretation carries with it a suggestion that the mobile shorelines of the Pleistocene were instrumental in bringing about this dissemination and subsequent divergence in isolation, just as they were for the tetraploid species of *Gossypium*. If this reasoning is sound, we have a second estimate of the amount of evolutionary divergence that can occur in a period of time up to a million years—that is, an amount of divergence that taxonomists generally recognize as closely related but essentially distinct species.

Finally, the geographic distribution of the two subgenera of *Cienfuegosia*, the one principally South American and the other African, may be seen as an example parallel to that of the subgenera of *Gossypium*. As in the latter group, the subgenera of *Cienfuegosia* probably diverged with the breakup of the Gondwana landmass as Africa and South America (or the landmasses that were to become these continents) began to separate from each other and from the other protocontinents. Thus, the divergence of the two subgenera may be dated as approximately of Cretaceous age. Again as in the case of *Gossypium*, one might argue either way for subgeneric or generic rank for these taxa, as I have discussed for *Cienfuegosia* in chapter 7. Thus, the level of divergence achieved in this time period (up to sixty-five million years) is almost to the level of what taxonomists recognize as distinct genera.

On Gossypol

IT seems inappropriate in a treatise on the Gossypieae to omit any discussion of the most distinctive and unifying feature of the group—the occurrence of gossypol, the unique sesquiterpenelike compound (or group of related compounds) that is characteristic of these plants. As yet, beyond purely technical considerations (molecular structure, commercial and nutritional importance, and so on), there is relatively little to be said about its evolutionary significance or about its role in the biology of the plant.

Most of the considerable attention that has been paid to this substance has been concerned with its undesirable properties, either as a discoloring contaminant in the oil extracted commercially from cottonseed or as a toxic contaminant in the protein-rich meal remaining after oil extraction. The great commercial importance of cottonseed oil and cottonseed meal has insured the gossypol receive thorough study by chemists, food technologists, and nutritionists. Consequently, we know a good deal about these aspects of gossypol.

But its role in the plant is less well understood, and its potential usefulness to man has not been fully explored, it seems. The native Hawaiians stripped the bark from the *Kokia* trees that are (or were) endemic to Hawaii and extracted from it a red dye, which they used as a preservative on their fishnets. It seems probable that this dye was a gossypol-rich substance and that the gossypol was the active agent in its preservative action. It has recently been shown (Gaind and Bapna, 1967) that the extract of green fruits of *Thespesia populnea* has strong bactericidal properties. This species is known to be rich in gossypol, and the extraction technique used was one that is appropriate for the isolation and concentration of gossypol. It seems probable that the bactericidal property of *Thespesia* fruit extract and the preservative quality of *Kokia* bark extract are measures of the same thing. A recent report (Vichkanova and Goryunova, 1968) of the viricidal activity of gossypol that has not to my knowledge been followed up is extremely intriguing. There is also anecdotal information available that various members of the Gossypieae (*Cienfuegosia, Thespesia*) are used medicinally; leaf decoctions are reportedly drunk by certain primitive peoples for various ills, but no evidence is available on the efficacy of these remedies or on the possible role that gossypol may play in them. Yet all together there is an indication in this information of a possible useful role for gossypol, a substance that commercially is generally regarded as an undesirable waste product.

But what of the role of gossypol in the plant? This biochemical characteristic of its production is certainly firmly fixed in the genetic architecture of the tribe. It evidently occurs universally within the tribe and is unique to it, being unknown in other members of the Malvaceae; indeed, it is unknown elsewhere in the angiosperms. Its occurrence is uniquely correlated with the presence of distinctive "oil glands" (better, "gossypol glands") that are found more or less throughout the above-ground parts of the plant. These structures are approximately spherical lysigenous glands that are filled with a mixture of pigments, of which gossypol is the dominant constituent. Pure gossypol is yellow. The whole glands vary in color from a translucent yellow (for example, in *Gossypium thurberi*) through various darker shades of red and purple to an essentially black color (as in the cultivated *Gossypium hirsutum*). It is tempting to suppose that the yellow glands contain relatively pure gossypol, uncontaminated with anthocyanins and other pigments, but I do not believe this is established as fact. The glands seem to serve as physiological "dumping grounds" for metabolic by-products, since they are cut off from metabolic interchange by suberization after they have been formed. This situation makes it difficult to speculate on the metabolic role of gossypol within the plant.

However, not all of the gossypol occurs in the gossypol glands. It also occurs in the root system, where it is synthesized and where glands are not developed. It also occurs in the "glandless" strains of *Gossypium hirsutum* that have been developed in recent years by plant breeders and which have proved to be very useful experimental tools. The story of these strains merits a special telling.

The commercial importance of gossypol (that is, its undesirability in cottonseed oil) suggested the possibility of breeding cottons that were free of this substance. Since the presence of gossypol is correlated with the presence of glands, it was feasible to approach the problem by seeking gland-free cotton plants. Intensive breeding efforts and genetic studies, initiated by the work of Scott McMichael, eventually gave rise to strains of cotton (*Gossypium hirsutum*) that are completely free of glands. Certain of these strains are in limited commercial production today. It should be noted that this concerted effort of plant breeding has produced a phenotypic expression (glandlessness) that not only is new in the species mentioned but also is previously unknown in the entire genus and indeed in the tribe. I do not wish to suggest that plant breeders are omnipotent; I will only make the more limited claim that they can sometimes do what appears on the surface to be impossible.

These glandless cottons show a reduced amount of gossypol in their tissues. It is, indeed, reduced to a very low level, but it is not absent. However, the level of gossypol in the embryo is sufficiently low that from the point of view of commercial oil extraction it is essentially absent. Thus, these cottons are successful as far as the commercial seed crusher is concerned.

Experience at growing these glandless cottons provides some possible insight into the biological role of the substance. Several observers report glandless cottons to be attacked by certain insects that normally do not occur on cotton. It is as if these insects, such as darkling beetles and lygus bugs, are able to feed on a cotton plant to a damaging extent only after the deterrent effect of the gossypol glands has been removed. This has not proved to be an insuperable problem in growing glandless cotton, because controlling insect pests is simply an accepted part of modern agricultural practice. However, it suggests that the presence of glands and of gossypol is a deterrent to insect depredation and that protection from damaging insects is the adaptational basis for the presence of these characteristics of the tribe Gossypieae.

In support of this idea is the finding that under experimental conditions an increased level of gossypol in the diet of certain cotton insects (*Heliothis, Anthonomus*) results in a reduced growth rate and higher mortality of larvae, which suggests that there is not only a deterrent effect of gossypol upon these insects but a toxic effect as well. Consequently, we are treated to the vision of certain cotton breeders dedicating their labors to increasing the amount of gossypol in the cotton plant while others are working equally hard to remove it altogether. The goals are clearly incompatible, and it is impossible at this writing to predict the gossypol content of the cotton plant of the future.

But it is difficult to accept the idea that gossypol functions as a toxin to protect the plants from insect pests. For one thing, many of the insects that are feeders on cotton plants are highly host specific. The boll weevil, for example, finds satisfactory hosts only within the tribe Gossypieae—primarily on *Gossypium hirsutum* and *G. thurberi*; also on *Cienfuegosia affinis, C. rosei,* and *Hampea nutricia*; and to a lesser extent on other species (Cross et al., 1975). The lepidopteran genus *Diparopsis*, a cotton pest of Africa, consists of four species, each of which has a somewhat different host range but which collectively are confined to *Gossypium Cienfuegosia,* and *Gossypioides* as their natural hosts (Galichet, 1964). If gossypol were indeed a toxin with a specific protective role to play in guarding the plant from insect feeders, one would not expect to find (1) that certain insects were especially adapted to feeding on these plants, or (2) that cotton (or any other members of the Gossypieae) had any insect pests at all. There are a number of insect species that are so adapted, and the significance of insect pests generally on the commercial cotton crop is attested to by the millions of dollars spent annually on insecticides and on the salaries of cotton entomologists! No, gossypol is toxic to insects only to the extent that almost any substance is toxic if consumed in large enough quantity.

Moreover, if gossypol had a role to play in the life of the plant as a protective device against insects, whether as a deterrent or as a toxic agent, one would not expect it to be so neatly isolated in suberized packages (the gossypol glands). These glands could easily be avoided by a feeding insect, if the insect found the contents of the gland to be distasteful or otherwise objectionable. I am not suggesting that a cotton leafworm larva has sufficient intelligence to discriminate between the gossypol-rich gland and the gossypol-poor (or -free?) mesophyll. I am only suggesting that mechanically it is possible for the larva to feed on the latter and avoid the former, in terms of the relative sizes of the insect's mouth parts and of the glands. If the plant had evolved gossypol and the gossypol glands as a defense against chewing insects, it is in the nature of coevolutionary relationships for the insects in their turn to have responded with evolutionary adaptations that would accommodate to the first change. So obvious a behavioral adaptation as gland avoidance would hardly have been overlooked as an evolutionary ploy. Yet none of the many insects that feed upon cotton, to my knowledge, exhibits this behavioral pattern. I thus conclude that the metabolic role of gossypol in these plants is not closely related to protecting them from insects.

I believe that the solution to this applied problem is a compromise. The purpose behind developing glandless varieties of cotton and behind the extensive promotional activities that are being used to "sell" these varieties to the cotton industry is simply to remove the glands (and the gossypol) from the seeds or , more specifically, from the embryos, not from the entire plant. If an increased level of gossypol in the leaves, buds, and other flower parts indeed has a role to play in the biological control of cotton insects, it would be desirable to manipulate the genotype in such a way as to produce a plant with gland-free embryos but with an increased level of gossypol in the rest of the normally glanded parts of the plant. It is known that the distribution of these glands to different parts of the plant is variable from species to species (and is thus under genetic control), and specifically that the desired combination—glands absent only from the embryos—occurs in several species of *Cienfuegosia* and in at least two species of *Gossypium* (*G. australe* and *G. bickii*). That is to say, it is biologically possible. Thus, if the breeders are able to produce what has never existed before (a glandless cotton plant), they should certainly be able to duplicate a genic configuration that is already known to exist (a glanded plant with gland-free embryos). They have only to bring it about.

We are left, then, without having accounted for the adaptational basis for the capacity to synthesize gossypol that is unique to the members of the Gossypieae or for the remarkable structures, the gossypol glands, that occur in conjunction with this capacity. Recent work by A. A. Bell and collaborators suggests, however, a possible basis. Using the glandless varieties of cotton that have been made available in recent years, they have shown that these plants are capable of being induced to produce a relatively high level of gossypol by certain kinds of stress. The kinds of stress that were imposed included disease organisms and chemical agents. A role of gossypol as phytoalexin, protective agent against fungal disease, was proposed and has been substantiated in subsequent studies. Such a role is consistent with the finding of antimicrobial activity for gossypol that was noted earlier and in total context is entirely plausible. It would have been impossible to arrive at this view, however, without the use of the gland-free strains of cotton for the experimentation.

9

Primitive and Advanced Traits

The pseudo-science of phylogeny . . . only diverts taxonomy into a bastard activity between science and fiction.
Lloyd H. Shinners, 1962

In spite of the hazards and uncertainties involved, the construction of phylogenies is a worthwhile occupation because it gives insight into relationships between organisms.
G. Ledyard Stebbins, Jr., 1974

Certain structural changes have occurred repeatedly in the evolution of flowering plants. Evidence for these evolutionary trends is often in the form of character gradients or morphoclines, within which the direction of change can be inferred from paleobotanical, geographical, or ecological circumstances.
Richard Eyde, 1971

THE phylogenetic views that now pervade systematics and which indeed have provided the rationale for the progression from the earlier artifical systems to the still-developing natural systems require that there be a conceptual scale relating the changing forms of characters, as they are expressed in living organisms. Ideally, this is a change from simple to complex: the simple forms are found in the older, ancestral organisms; the complex forms are found in the younger, derivative organisms. In practice the change is rarely this simple.

Some of the qualifications that must be considered (none of which can be dealt with in detail here) are the following: (1) the fossil record is highly imperfect, especially for vascular plants, so it is generally difficult to relate different character states directly to a time scale; (2) evolution is not unidirectional, so that whereas change often occurs from simple to complex, reversion also occurs, giving rise to change in the opposite direction; (3) rates of evolution are not constant, either in comparing different groups of organisms or in comparing different characters in the same organism; (4) it is not always (or even often) self-evident which character states are primitive and which are derived, especially in the absence of fossil evidence; and (5) phylogenetic history is not always linear but may be convergent or reticulate.

In spite of these difficulties, much useful work has been done, and general consensus has been reached concerning the nature of primitive and advanced characters in phanerogams (cf. Davis and Heywood, 1963, pp. 34 ff.; Eyde, 1971). In a way it is surprising that this agreement should have been attained, because the context of the subject is always potentially open to the danger of circular reasoning. The danger of circularity is diminished, however, when the concepts of primitive and advanced characters are derived from or applied to studies of a single higher taxon or phyletic lineage. A more intimate knowledge of the organisms involved usually enables satisfactory biological answers to be give to questions that, in a logical sense, are inherently circular (cf. Estabrook, 1977). It is at this level that the critics of phylogenetic taxonomy can be answered. The present chapter is an exercise in this genre.

We are faced, in the Gossypieae, with an essentially blank fossil record. We must infer from extant organisms everything that can be said about the evolutionary history of the group. There are those who would say that in the absence of a fossil record nothing of value can be inferred; some would go even further and assert that it is positively undesirable to "speculate" on phylogenetic history in the absence of a fossil record since such "speculation" may be misleading. Though there is an element of truth to this view at the formal, logical level, the criticism can generally be answered in specific instances from biological considerations. From one viewpoint the question becomes: What qualifies to be called "speculation"? To me, *speculation* may be defined as "reasoning unsupported by evidence"; to others, evidently, it is defined as "reasoning unsupported by fossil evidence." Certainly fossil evidence, where available, takes first place in considerations of phylogenetic history, but where it is not available, inquiry need not cease. Other kinds of evidence can be very compelling, especially when buttressed by a comprehensive knowledge of the biology of the group.

Two kinds of inquiry may be followed. On the one hand, the characters that are common to the members of the tribe may be evaluated for their primitive or advanced condition in an effort to characterize the group with respect to its relative advancement and to relate it to the remainder of the vascular plants, especially its immediate relatives. On the other hand (and perhaps more fruitfully), the characters that vary among the members of the tribe may be considered for their relative advancement for the light this may shed on the phylogenetic history of the members of the group.

The first type of inquiry may be conducted quite briefly. Primitive characters that are common to members of the Gossypieae are the tropical distribution; the perennial habit; spirally arranged leaves that are usually simple and stipulate; large, polypetalous, actinomorphic flowers; hypogyny; pleiomerous stamens; and epigeal germination. Advanced characters that are common to the members of the Gossypieae include palmately veined leaves, stellate and lepidote indumentum, androecium specialized by connation of filaments into a staminal column, pentamerous (rather than pleiomerous) calyx and corolla, syncarpous fruits, multiaperturate pollen, well-developed extrafloral nectaries, reduced endosperm (at maturity of seed), and conduplicate cotyledons and a large, complex embryo.

Thus, we find that the Gossypieae are a mixture of primitive and derived characters, as are most groups of plants. The Malvaceae as a whole (indeed, the order Malvales) are generally placed among the relatively more primitive families of the dicots (as by Sporne, 1954). It is significant to note that of the primitive characters shared by members of the Gossypieae, most are also shared with the Malvaceae as a whole; of the advanced characters shared by members of the Gossypieae, many of them (including syncarpous fruits, pollen, nectaries, and embryo) are *not* shared with the Malvaceae as a whole. The tribe Gossypieae thus is relatively specialized within the family in which it is included.

Within the tribe, the genera show varying combinations of primitive and advanced characters, as shown in Table 4. The entries in that table are to a degree arbitrary, and a comment on the arbitrariness is in order. It is perhaps a subjective matter, and certainly a relative one, whether flowers are "large and showy" or "small and inconspicuous." For example, *Gossypium* generally has showy flowers (and is so categorized in the table), but sometimes, as in *G. somalense*, it has inconspicuous flowers. Conversely, the flowers of *Hampea* are generally quite small and inconspicuous, but those of *H. latifolia* and *H. breedlovei* are relatively showy. The flowers of the monotypic *Lebronnecia* are intermediate on this scale of size and might be categorized either way, depending upon how the categories are (arbitrarily) distinguished. Similar comments might be applied to several of the other character distinctions in Table 4. The data in Table 4 have been used to derive the similarity indices presented in Table 5, which gives the proportion of character states in common for all possible pairs of taxa.

With these and similar qualifications in mind, and ever mindful of the limitations inherent in such data, we find it is nevertheless possible to draw some conclusions concerning the relationships among these taxa. From Table 4 we may conclude that *Thespesia* and *Cephalohibiscus*, closely followed by *Kokia*, are the least specialized genera in the tribe and that the other genera are relatively more specialized. It is probably not meaningful to attempt to make finer distinctions on the basis of these data.

From the correlation patterns of Table 5 we may conclude that *Cienfuegosia* is the genus most isolated from the remainder of the genera because of its distinctly lower average similarity index. *Hampea* is the next most isolated genus of the group. The remainder show relatively high degrees of affinity. *Lebronnecia*, with the highest average similarity index, shows a central position in the group. Looking at the genera individually, we find that *Thespesia* has its closest affinity with *Cephalohibiscus* and that the reciprocal relationship also holds. This is consistent with van Borssum Waalkes' view that these plants are congeneric. *Kokia* has its strongest relationship with *Cephalohibiscus*, thus fitting into the same group. *Lebronnecia* and *Hampea* show a strong mutual affinity (the highest value in the table), and *Gossypium* and *Gossypioides* likewise show their strongest affinity for each other. All of the preceding observations are based on similarity indices of 0.74 or higher. The highest value shown by *Cienfuegosia* (0.58 when compared with *Gossypium*) is markedly lower, which emphasizes the relatively isolated position of *Cienfuegosia* in the Gossypieae.

These conclusions concerning affinity among the genera are, by and large, ones that would be reached in a more subjective manner by examination of the plants. The quantification provided by the similarity indices strengthens the conclusions. It would be foolhardy in the present state of knowledge to go beyond these general statements to attempt to propose any inferred phylogeny among the genera.

It is tempting, however, to use the above conclusions to extrapolate and speculatively to "predict" the chromosome numbers of the two genera of the Gossypieae whose chromosome numbers have not been determined. Both *Cephalohibiscus* and *Lebronnecia* "should" have chromosome numbers of $2n = 26$ on the basis of their marked affinities to *Thespesia* and *Hampea*, respectively (cf. chapter 7).

Table 4. Primitive and Advanced Character States
of the Genera of Gossypieae (o = primitive; + = advanced).

Character State	Gossypium	Thespesia
woody (o) vs. herbaceous (+)	o	o
trees (o) vs. shrubs (+)	+	o
flowers dimorphic (o) vs. monomorphic (+)	o	o
fruits pentamerous (o) vs. trimerous (+)	+	o
seeds pubescent (o) vs. glabrous (+)	o(+)	o
calyx 5-lobed (o) vs. truncate (+)	+	+
seeds arillate (+)	o(+)	o
mesophytic (o) vs. xerophytic (+)	+	o
involucel ∞ (o) vs. 3-parted (+)	+	o
involucel bracts subulate (o) vs. foliose (+)	+	o
flowers showy (o) vs. inconspicuous (+)	o	o
pantropical (o) vs. localized (+)	o	o
foliar nectaries (+)	o	o
embryo conduplicate (+)	+	+
chromosome number $2n = 20, 22,$ or 24 (+)	o	o
capsules pubescent (o) vs. glabrous (+)	+	o
stipules foliose (+)	o	o
flowers solitary (o) vs. sympodial (+)	+	o
calyx obscurely punctate (o) vs. prominently punctate (+)	+	o
Number of advanced traits	10	2

It is difficult to reach firm conclusions on generic relationship from a consideration of modal character expression of the several genera. Nevertheless it is possible to suggest certain affinities from a consideration of trends of variation *within* the genera. Certain of these trends and their implications are discussed below.

In *Gossypium* shrubbiness is a modal expression for growth habit. Yet certain species are truly arborescent, as are the three species included in section *Erioxylum*: *G. aridum*, *G. lobatum*, and *G. laxum*. These species show a complex of other characters that may also be considered primitive and thus may be viewed as representing a transitional type between the more specialized representatives of *Gossypium* (or at least of its American representatives) and an earlier ancestral type from which the genus sprang. I am not, of course, suggesting that *G. aridum* is in itself ancestral to the balance of the genus, but only that its character combination more nearly resembles the character combination of such an unknown ancestor (Fig. 55). Viewed thus, it might also be considered as a potential link with other

Cienfuegosia	Hampea	Gossypioides	Cephalohibiscus	Lebronnecia	Kokia
+	o	o	o	o	o
+	o	+	o	o	o
o	+	o	o	+	o
+	+	+	o	+	o
o	+	o(+)	o	o	o
o	+	+	+	+	o
o	+	o	o	o	o
+	o	o	o	+	o
o	+	+	+	+	+
o	o	+	o	o	+
o	+	o	o	o or +?	o
o	+	+	+	+	+
+	o	o	o	o	o?
+	+	+	o	+	+
+	o	+			+
+(o)	o	o	o	o	+
+(o)	o	+	o	o	o
o	o	+	+	o	
+	+	+	+	o	+
10	10	11	4	8	6

genera of the tribe. It is through such an arborescent *Gossypium* that we might seek links with other arborescent genera, such as *Thespesia* and *Hampea*, in the tribe.

In *Hampea* the species of section *Standleya* (*H. platanifolia, H. latifolia,* and *H. rovirosae*), which are the only perfect-flowered species in the genus, are interpreted as having relatively the most primitive character expressions in *Hampea*. In addition to having perfect flowers, these species have relatively larger flowers, a greater number of seeds per locule, more highly developed involucral nectaries, and more primitive calyx morphology than the balance of the genus. Also, the number of flowers in each axillary fascicle is only one or two in *H. rovirosae* and *H. latifolia* in contrast to the denser fascicles found in other species (for example, *H. nutricia*), where as many as twelve flowers may occur in each axillary position. Thus, these more primitive representatives of section *Standleya* indicate an alliance with the other genera of the Gossypieae in which axillary flowers are generally solitary.

Table 5. Similarity Indices Derived from Table 4.

Genus	$\bar{x}$	Thespesia	Cephalo-hibiscus	Kokia	Lebron-necia	Gossypium	Gossypi-oides	Hampea	Cienfue-gosia
Thespesia	(.61)		$\frac{14}{18} = .78$	$\frac{11}{17} = .65$	$\frac{12}{18} = .67$	$\frac{11}{19} = .58$	$\frac{10}{19} = .53$	$\frac{11}{19} = .58$	$\frac{9}{19} = .47$
Cephalohibiscus	(.61)			$\frac{12}{16} = .75$	$\frac{12}{18} = .67$	$\frac{10}{18} = .56$	$\frac{12}{18} = .67$	$\frac{10}{18} = .56$	$\frac{5}{18} = .29$
Kokia	(.60)				$\frac{10}{16} = .63$	$\frac{11}{17} = .65$	$\frac{12}{17} = .70$	$\frac{8}{17} = .42$	$\frac{8}{17} = .42$
Lebronnecia	(.66)					$\frac{12}{18} = .67$	$\frac{12}{18} = .67$	$\frac{14}{17} = .82$	$\frac{9}{18} = .50$
Gossypium	(.61)						$\frac{14}{19} = .74$	$\frac{9}{19} = .47$	$\frac{11}{19} = .58$
Gossypioides	(.60)							$\frac{10}{19} = .53$	$\frac{10}{19} = .53$
Hampea	(.52)								$\frac{5}{19} = .26$
Cienfuegosia	(.44)								

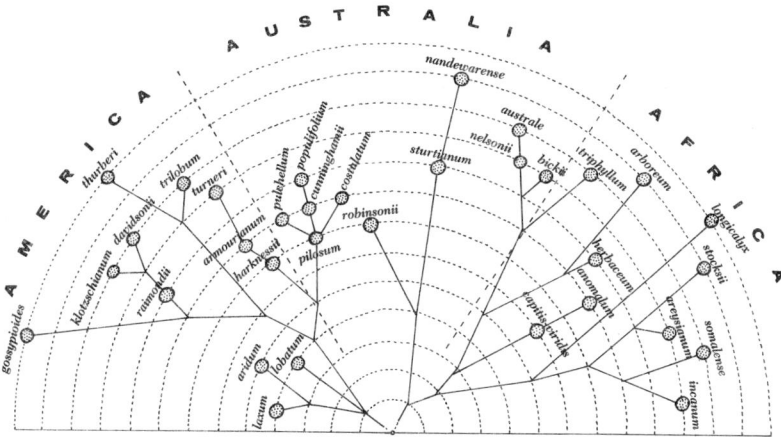

Fig. 55. Diagrammatic branching sequence for the phylogeny of the diploid species of *Gossypium*. (Adapted from Fryxell, 1971.)

In addition to axillary flowers, however, flowers are borne on sympodial inflorescences in *Lebronnecia*, in many species of *Gossypium*, in *Gossypioides*, and in some species of *Cienfuegosia*. It is thus of interest to find that *Hampea micrantha* has its flowers borne on a reduced, branched peduncle rather than in fascicles. It is impossible at this point to determine if the fasciculate inflorescence that is typical of *Hampea* was derived from the further reduction of a branched inflorescence of the type found in *H. micrantha* or whether it was derived by proliferation from a solitary-flowered pattern of the type found in *H. rovirosae*. Developmental studies may yield an answer to this question.

Another species of *Hampea* showing primitive characters is *H. tomentosa*, although it has been of somewhat uncertain systematic position in the genus. In many ways it shows many character expressions that are as primitive as those of the species of section *Standleya* (except that it has dimorphic rather than hermaphroditic flowers). It has highly developed involucral nectaries and a primitive calyx (five-ribbed, with a five-toothed margin), and it has a relatively woody capsule. Although the flowers are fully dimorphic, those of pistillate individuals do rarely produce a few pollen grains. Thus, to the extent that the flowers of pistillate trees are in fact pollen fertile, the species may be regarded as androdioecious instead of strictly dioecious. The pistil is completely suppressed in the flowers of staminate individuals, so the species is indeed dimorphic, but the occasional pollen fertility of the otherwise pistillate individuals indicates an intermediacy of this species between the perfect-flowered representatives of section *Standleya* and the dioecious species of the remainder of the genus. It seems likely that dioecism in *Hampea* was achieved via androdioecism.

The most striking feature of *H. tomentosa* has only recently come to light, however, and it provides significant evidence concerning generic alliances with *Hampea*. *H. tomentosa* was first described by Presl as a species of *Thespesia*. The involucel was originally described as follows: "Involucrum octophyllum, phyllis setaceis tomentosis sesquilineam longis fugacissimis in alabastro valde juvenili solummodo conspicuis ad basim calycis tres callositates transverse oblongas nigricantes reliquentibus." An eight-parted involucel would not be unusual for *Thespesia*, and Presl was justified in placing the species in that genus. (One must also note that the genus *Hampea* had not been described at the time Presl wrote but was described in the following year, based on another species.)

When I undertook to study *Hampea* and became familiar with the many species of *Hampea* now known with tripartite involucels, I was skeptical of Presl's description of a multipartite involucel, believing that it was either an erroneous observation or that the species had been placed in *Hampea* (by Standley) incorrectly. It was therefore with eagerness that I first examined the holotype, intent upon resolving this question. The holotype includes only a few flowers or buds. None of these had an involucel, the involucels having been shed at an earlier stage, but the very prominent nectaries were much in evidence, and the plant was certainly a *Hampea*. I therefore concluded that Standley had made the correct transfer and that Presl, for whatever reason, had made an incorrect observation. I was puzzled why this error should be so but could see no means of resolving the question. The only other material of *H. tomentosa* that was available to me at that time shed no further light on the matter.

When I located plants of *H. tomentosa* in the wild and had the opportunity for the first time to observe the species in flower, I found that Presl was vindicated. The very young buds *do* have an involucel, which is deciduous long before anthesis. Moreover, it is not a tripartite involucel, as in other species of *Hampea*, but a multipartite one. Above each of the three large nectaries is a group of very small, filiform bractlets. Each group usually consists of three bractlets (thus giving a total of nine bractlets per involucel), but occasionally there are fewer.

Because of their small size, slightly irregular number, and early loss, Presl may be forgiven for saying that the involucel is eight-parted rather than nine-parted. Evidently Presl based this observation either on field notes or on a part of the type material (an isotype) that was not available to me. In any case, his observation was correct.

Upon discovering this primitive feature of *H. tomentosa*, which I had not previously suspected to exist in *Hampea*, I examined a specimen of *H. rovirosae* that I was growing in greenhouse cultivation. The involucral bractlets of this species are also very small and deciduous before anthesis. The involucral nectaries, on the other hand, are very prominent. The three nectaries of *H. rovirosae* are very irregularly positioned (a characteristic also true of *Lebronnecia* and of some species of *Thespesia*) and are not whorled, as is typical of the tribe generally. Indeed, there is sometimes a fourth nectary on the pedicel of *H. rovirosae* well below the flower, suggesting that the normally whorled involucel of the Gossypieae was derived by foreshortening of structures that originally occurred in a spiral disposition. Typically, each nectary of *H. rovirosae* is surmounted by a small bractlet, and the involucel as a whole is thus tripartite. Rarely however, there are *two* small bractlets associated with a single nectary. I have since seen the same situation on young buds of *H. latifolia*, which supports an alliance of section *Standleya* as a whole with *H. tomentosa*.

What do these observations suggest? Throughout the tribe Gossypieae involucels are basically trimerous and commonly tripartite. In some *Thespesias* (as in *T. garckeana*) the multipartite involucel shows no indication of a trimerous nature; in other *Thespesias* (such as *T. lampas*) and some *Cienfuegosias* (such as *C. gerrardii*) a nine-parted involucel is clearly trimerous, with a group of three bractlets associated with each of three nectaries, exactly as in *Hampea tomentosa*. Since both *C. gerrardii* and *H. tomentosa* have a set of character states that may be regarded as relatively primitive within their respective genera, their common possession of a nine-parted involucel may be considered as suggestive of a possible alliance with each other and with that part of *Thespesia* having a similar involucel, exemplified by *T. lampas*.

Thus, although generic alliances in the Gossypieae are not obvious superficially, some evidence concerning the nature of these alliances can be gained from comparisons such as those indicated.

Graft Compatibilities

STUDIES of graft compatibilities and incompatibilities can potentially yield a class of data otherwise unavailable for evaluating relationships. This kind of data has been only partially explored for evaluating relations among the genera of the Gossypieae.

As far as we now know, no intergeneric hybridizations are possible within the tribe Gossypieae. Indeed, many intrageneric (interspecific) hybridizations are also impossible. It has been shown in other plants (for example, *Theobroma* and *Trifolium*) that evaluations of relationships among species, as measured by intergrafting experiments, are highly correlated with evaluations based on hybridization experiments. If graft compatibility experiments have greater "reach" than hybridization experiments (as they evidently do), it should be possible to extend our knowledge of genetic relationships among the taxa of the Gossypieae. For example, Beckett (1933) successfully grafted *Thespesia populnea* on four species of *Gossypium* (*G. sturtianum, G. hirsutum, G. aridum,* and *G. arboreum*), whereas to my knowledge these intergeneric hybridizations are impossible.

As yet, only limited data are available on intergeneric graft compatibilities in the Gossypieae. As just noted, Beckett has demonstrated compatibility between *Thespesia* and *Gossypium*. Hutchinson (1943) found that *Gossypioides brevilanatum* would graft easily on *Gossypioides kirkii* but failed to graft on four species of *Gossypium* (*G. aridum, G. arboreum, G. hirsutum,* and *G. barbadense*), thus supporting the taxonomic distinction of these genera in spite of their morphological resemblance. Hutchinson (1943) also reported that *Kokia rockii* and *Gossypioides kirkii* graft successfully but that translocation across the graft union is completely or almost completely blocked, indicating that the compatibility is marginal, at best, and that these two genera are not closely related.

Additional studies of graft compatibility, some of which are under way, involving other genera of the tribe should yield additional insights into generic relationships that are impossible to achieve from hybridization experiments or studies of comparative morphology.

10

Overview

*In all branches of natural history—Astronomy, Geology,
Meteorology, Botany and Zoology—classification, whether it is
of stellar spectra, of igneous rocks, of pressure-distributions, of
algae or of worms, has always occupied a central position,
comparable perhaps to that which natural philosophers accord
to mathematics.*

R. A. Crowson, 1970

*The biologist who dares to ignore the provincial boundaries can,
with broad training and insight, approach any of the limitless
interfaces between ecology, systematics, and evolution. Then he
can confront effectively the most fascinating problems of all—
why organisms are what they are, or why they do what they do.
In a nutshell, superintegration of the three fields of synthesis
truly will give us a twentieth-century natural history.*

A. R. Kruckeberg, 1969

Parallel Evolution

ONE of the striking facets to emerge from this review of the systematics, comparative morphology, and evolutionary trends exhibited by the Gossypieae is the frequency with which these trends cut across taxonomic groupings instead of following them. A superficial response might be to conclude that the taxonomic groupings are artificial and in need of major revision. Even a modest familiarity with the plants, however, shows that the systematic groupings presented in chapter 2 are basically sound and that such a suggestion for revision is indeed superficial.

There are instances, of course, in which certain characteristics and trends follow the accepted taxonomic divisions, including the herbaceous habit confined to *Cienfuegosia*, the whole pattern of floral biology and seed dissemination that is characteristic of *Hampea*, and the pattern of chromosome numbers for the tribe in which the different numbers follow generic or infrageneric boundaries.

In many other cases, however, trends cut across taxonomic groupings, and we must speak of parallel evolution. Evidently the several evolutionary lineages, treated here in taxonomic terms as genera, sub-genera, and sections, have retained enough evolutionary plasticity to respond similarly but separately to similar evolutionary challenges and opportunities. Thus, for example, we find the typical hairy seeds of *Cienfuegosia* modified to the tightly appressed condition in *C. drummondii* and to the subglabrous condition in *C. affinis*. Similarly, in *Gossypium* we find the typical hairy seeds modified to the tightly appressed condition in *G. harknessii* and to the subglabrous condition in *G. trilobum*. We find narrowly funnelform, purple corollas in *Cienfuegosia hearnii* and in *Gossypium aridum*; we find open, yellow, spotless corollas in *Cienfuegosia yucatanensis* and *Gossypium tomentosum*; we find very large, reddish corollas in *Thespesia grandiflora* and *Kokia kauaiensis*; and we find distinctively exserted androecia in *Thespesia cubensis* and *Cephalohibiscus peekelii*. But these are all parallel developments of specializations within different lineages and are not indicative of close relationship, except in the sense that these plants (or their ancestors) have similar ranges of genetic variability and the ability to respond similarly to selection pressures and to evolve parallel phenotypes.

The conclusion seems inescapable that those differences which follow generic (and infrageneric) lines, and thus distinguish the different lineages, occurred relatively earlier in evolutionary history and that those differences which cut across taxonomic groupings occurred relatively more recently in evolutionary history, after the several lineages were established. How do we tell the difference? This, I suppose, is a fundamental question for systematics that deserves a categorical answer, but I doubt that it will receive one. Or at least such an answer will have to be couched in relative terms and make ample reference to the touchstone of systematics, the "unending synthesis" (Constance, 1964) of all available data, rather than the emphasis of single criteria. It is from such a consideration of the totality of the information that we draw conclusions about the boundaries of genera and subgenera and about the naturalness of the groups so delimited. It is only after we have arrived at these conclusions that we can focus our attention on those characters that cut across the taxonomic groups we have perceived and that we can consider as a phenomenon (parallel evolution) superimposed on the phyletic pattern of the different lineages.

When we do this in the Gossypieae, what do we see? We see that many of the parallels involve characters related to reproductive biology:

methods of attracting different classes of pollinators, adaptations toward facultative inbreeding or facultative outbreeding, and adaptations for different methods of seed dispersal. Evidently such characters retain considerable evolutionary plasticity and can respond to appropriate selection pressures on a relatively short-term basis. This accords well with what we know of other plant groups. I think immediately of examples such as *Hibiscus poeppigii* (Sprengel) Garcke, *Malvaviscus arboreus* Cav., and *Periptera punicea* (Lag.) DC., referred to in chapter 8, all of which have evolved the same highly specialized floral phenotype (narrow, tubular, bright red corollas with long-exserted anthers and stigmas, adapted to hummingbird pollination) in spite of representing not only three different genera but three different tribes of the Malvaceae. Other examples might be presented, but it is tempting to suggest that perhaps the selection pressures are sufficiently intense to give rise to examples of parallel evolution more frequently when they result from coevolutionary relationships (in this case with pollinators) than when other kinds of selection pressure are involved.

The Evolutionary Continuum

SYSTEMATISTS are vulnerable to the criticism that they often fail adequately to take into account the time dimension when constructing their systems. This criticism is particularly true at the philosophical or conceptual level, where the most basic concept of systematic biology must be dealt with: the species concept. Developing and conceptualizing a satisfactory species concept is a difficult enough task when considering only contemporary organisms, but when the time dimension is added through the consideration of the fossil record, difficulties are multiplied.

Consider the much-touted "biological species concept" (a redundancy if I ever heard one!), which, in oversimplified terms, defines a species as a potentially interbreeding population. Although it has many difficulties, there is much to be said for this definition. But when we project this definition to biological lineages evolving over geologic time, it becomes almost circular in its logic. Organisms separated by a million years are the same species because they are potentially interbreeding or are potentially interbreeding because they are the same species. What are the limits of "potential interbreeding"? At what level of divergence do diverging taxa or incipient taxa become distinct? How can we relate such concepts and phenomena to the established taxonomic categories?

These are the questions that must be dealt with at the conceptual level, and when the time factor is specifically included in one's thinking, the questions become difficult ones indeed.

Palaeontologists, of course, are very much aware of these problems, and many of them are insistent, even importunate, in forcing them to the attention of systematists dealing principally with contemporary organisms. Many of us wish they would go away, or at least be less insistent, because the problems they point out are difficult ones to contend with. But we should be grateful for their insistence, because the problems are of basic importance, and we cannot make sense of the biological world if we ignore basic problems.

I do not propose to discover a new Rosetta Stone to solve the conceptual problems of the evolutionary continuum. But I do wish to suggest a point of view that may help relate the evolutionary continuum to applied problems of classifying contemporary taxa.

If we think of the phyletic pattern of any segment (say, an order) of the biological world in terms of a dendritic pattern or "evolutionary tree," which is doubtless an oversimplification but satisfactory for present purposes, then we can think of a cross-section of this tree as a representation of those particular taxa to be found at a given time, such as the present. If the tree is taken to represent an order, the main branches that diverge near the base can represent families; a higher order of branching, tribes; above that, genera; then subgenera and species. Of course there will be uneven branching. All families will not diverge at the same level on the time axis, nor will tribes, nor genera. Therefore, an arbitrary cross-section in time will catch different lineages in different stages of differentiation, that is, nearer branching points in some cases and more distant in others. What does this mean in practical terms?

It has sometimes been noted that clearly delimited families have poorly delimited genera or vice versa, or that sharply delimited genera have poorly delimited species or the opposite. These have been referred to in a somewhat different context as "definable" and "indefinable" families by Walters (1961) and discussed by Davis and Heywood (1963, p. 107). I would not wish to propose that successive taxonomic ranks in a given lineage necessarily alternate in this respect, but there may be a tendency for this to be true. To the extent that it is so, the rationale of the evolutionary continuum makes it plausible.

If the present time intersects the phyletic tree shortly after a taxon has diverged but before it has fully differentiated into constituent taxa, we may expect to find a relatively clearly defined group with relatively poorly defined constituents. This situation is evidently what we find in the genus *Hampea* or in *Cienfuegosia* sect. *Cienfuegosia*, both of which are themselves clearly defined but include somewhat poorly defined species.

On the other hand, if the present time intersects the phyletic tree long after a taxon has diverged, we may experience difficulty defining it, but its constituents will be fully differentiated and clearly delineated. Such is the case with the genera *Gossypium* and *Thespesia*, both of which have been fragmented in the past into infrageneric taxa or segregate genera, the ranks of which have been debatable but whose constituent species are noncontroversial in their delimitation. We have simply caught *Gossypium* and *Hampea* in the present time at different evolutionary stages; the former genus is older, the latter younger. This is hardly a profound conclusion, yet if we recognize this reciprocal relationship as one that sometimes occurs in taxa at successive ranks, I think it will enable us to resolve taxonomic problems more effectively and to propose evolutionary interpretations more plausibly.

Synergistic Aspects of "Natural History"

IN the Introduction I dealt briefly with the question What is a natural history? The remainder of this work has been an exposition of the natural history of a particular group, the Gossypieae or cotton tribe. Now I wish to address the question of what advantages ensue from this broadly based orientation.

It is my contention that by bringing diverse viewpoints together and focusing them on a common group of organisms, the resultant understandings are synergistic and reinforce each other. By considering (for example) systematics and ecology, comparative morphology and pollination biology, population genetics and ethnobotany, we find that each field draws from the other, and the resultant understandings are more than the sum of the several contributions. We not only broaden but also deepen our knowledge. Scholarship of any kind is eclectic and synthetic in large measure.

Therefore, I find it incomprehensible and even a little painful when I hear respected scholars taking potshots at their colleagues in other fields and saying or implying that the other fellow's work is not really "scientific" or the other person's field is not really "rigorous." Often the critic does not understand the colleague's field (which makes it much easier to criticize) or he has encountered poor work done in another discipline and confuses the discipline with an individual's substandard work. Sometimes another field is criticized simply because it is different and thus suspect. To criticize is a much easier course than to understand, but it contributes nothing to the solution of scientific problems.

When a taxonomist says, "The pseudo-science of phylogeny . . . only diverts taxonomy into a bastard activity between science and fiction," has he done anything but alienate students of evolution from himself personally and from taxonomy as a discipline? When students of experimental evolution make slighting remarks about "museum taxonomists" and when cytologists decry a reliance on "superficial morphology" as an unworthy method of assessing relationships, is biological understanding advanced? I think not.

Science operates on many levels simultaneously. When biologists are faced with the problem of understanding the manifold nature of biological diversity, they must attack it at many different levels. The ecologist is concerned with variability within and among communities; the taxonomist with variation within and among families, genera, species, and other taxonomic units; the geneticist with variation within and among interbreeding populations or among individuals; and the developmental embryologist with variation among different organs and tissues. None of these fields of specialization is inherently superior to the others. All of them, to contribute meaningfully to the advancement of knowledge, must operate in a context of understanding of and interaction with the other disciplines. They are mutually supportive, at the least, and their interactions can be positively creative at best. These possibilities can be destroyed, however, by continuing guerrilla warfare, as when it is argued that the "evolutionary species concept" is superior to the "biological species concept," as though these were different biological phenomena instead of differences in words.

Taxonomists have been accused of causing trouble by working with cultivated plants and producing "messy classifications" and "over-classifying." Such criticism often comes from the ranks of geneticists, some of whom have a poor or oversimplified understanding of systematics. Cultivated plants are admittedly taxonomically difficult. This is a fact long known. It makes little sense to blame this fact upon those who have attempted to improve the taxonomy of these groups.

Taxonomic classification schemes of cultivated groups are often "poor," "messy," or "useless." That this is so has resulted as often from the work of "classical taxonomists" who are not adequately familiar with the plants as cultigens as it has from the work of geneticists and other nontaxonomists who, while familiar with the plants, have attempted to classify them in ignorance of the methods and objectives of systematics. To attempt to affix "blame" is sterile. What is needed is a renewed attack on the problem of classifying these difficult groups satisfactorily, new kinds of information, and new insights into relationships, because satisfactory classifications of these difficult groups are necessary to our understanding of the plants and to our ability to communicate about them. Without such classifications we are in the dark.

My point is that natural history, as an integrative summation of information from a variety of disciplines, whether these are descriptive, experimental, or historical in nature, can give rise to a richer and deeper understanding of biological diversity than can a more one-sided approach based only on descriptive studies or only on experimental studies. Another way to make the same point is to refer to the mutual interdependence of taxonomic and evolutionary studies. Most experimental evolutionary studies are concerned in one way or another with the *mechanics* of evolution at the specific level or below, expressed largely in genetic terms. Taxonomic studies, on the other hand, are mostly concerned with the *products* of evolution—species and higher taxa. Genetics and taxonomy are thus opposite sides of the evolutionary coin, the one concerned with mechanisms in an operative sense, the other with end products in a historical sense. Taxonomy, to be meaningful and to achieve its goal of a natural classification, must function in a frame of reference that is basically phylogenetic or evolutionary. Evolutionary biology, to achieve permanence of its results, must store its findings in a taxonomic system, the best method yet devised for the storage and retrieval of biological information. Taxonomy and evolutionary biology are indeed mutually interdependent.

Bibliography

Alefeld, F. G. C. 1861. Ueber die Stellung der Gattung *Gossypium* und mehrer Anderer. *Bot. Zeit.* 19:229–301.

Aliotta, A. 1903. *Rivista critica del genere* Gossypium. Portici: Della Torre. 111 pp.

Baker, E. G. 1897. Notes on *Thespesia*. *J. Bot.* 35:50–54.

Baker, H. G. 1961. The adaptation of flowering plants to nocturnal and crepuscular pollinators. *Quart. Rev. Biol.* 36:64–73.

Beasley, J. O. 1940*a*. The production of polyploids in *Gossypium. J. Hered.* 31:39–48.

————. 1940*b*. The origin of American tetraploid *Gossypium* species. *Amer. Nat.* 74:285–286.

Beckett, R. E. 1933. *Grafting Experiments with Cotton.* USDA Circular 267. 14 pp.

Bentham, G., and J. D. Hooker. 1867. Malvaceae. In *Genera Plantarum.* 3 vols. London (1862–1883).

Blanchard, O. J., Jr. 1978. An additional species, a new section, and an earlier epithet in *Cienfuegosia* Cav. (Malvaceae). *Ann. Missouri Bot. Gard.* 65:764–766.

Blanco, F. M. 1877–1880. *Flora de Filipinas*, 3rd ed. 4 vols. Manila.

Bolkhovskikh, Z., V. Grif, T. Matvajeva, and O. Zakharyeva. 1969. *Chromosome Numbers of Flowering Plants.* Leningrad: Acad. Sci. U.S.S.R., 926 pp.

Borssum Waalkes, J. van. 1966. Malesian Malvaceae revised. *Blumea* 14:1–213.

Capuron, R. 1968. Contributions à l'étude de la flore forestière de Madagascar: Un *Thespesia* nouveau de Madagascar (Malvacées). *Adansonia* n.s. 8:5–9.

Carlquist, S. 1974. *Island Biology.* New York: Columbia University Press. ix + 660 pp.

Chowdhury, K. A., and G. M. Buth. 1971. Cotton seeds from the Neolithic in Egyptian Nubia and the origin of Old World Cotton. *Linn. Soc. London, Biol. J.* 3:303–312.

Constance, L. 1964. Systematic botany—an unending synthesis. *Taxon* 13:257-273.

Corner, E. J. H. 1949. The Durian Theory or the origin of the modern tree. *Ann. Bot.* (London) n.s. 13:367–414.

————. 1954*a*. The Durian Theory extended I. *Phytomorphology* 3:465–476.

————. 1954*b* The Durian Theory extended II: The arillate fruit and the compound leaf. *Phytomorphology* 4:152–165.

————. 1976. *The Seeds of Dicotyledons.* 2 vols. New York: Cambridge University Press.

Cronquist, A. 1968. *The Evolution and Classification of Flowering Plants.* Boston: Houghton-Mifflin. xi + 396 pp.

Cross, W. H., M. J. Lukefahr, P. A. Fryxell, and H. R. Burke. 1975. Host plants of the boll weevil. *Environ. Entomol.* 4:19–26.

D'Arcy, W. G. 1976. Near extinct plant in climatron. *Missouri Bot. Gard. Bull.* 64 (3).

Davis, P. H., and V. H. Heywood. 1963. *Principles of Angiosperm Taxonomy.* New York: Van Nostrand. xx + 558 pp.

Degener, O. 1932–1965. Kokia. In *New Illustrated Flora of the Hawaiian Islands.* Fam. 221.

Douwes, H. 1953. The cytological relationships of *Gossypium areysianum* Deflers. *J. Genet.* 51:611–624.

Dumont, A. 1887. Recherches sur l'anatomie comparée des Malvacées, Bombacacées, Tiliacées, Sterculiacées. *Ann. Sci. Nat.* ser. 7, 6:129–246.

Edlin, H. L. 1935. A critical revision of certain taxonomic groups of the Malvales. *New Phytol.* 34:1–20, 122–143.

Edwards, G. A. 1977. The karyotype of *Gossypium herbaceum* L. *Caryologia* 30(3):369–374.

Endress, P. K. 1973. Arils and aril-like structures in woody Ranales. *New Phytol.* 72:1159–1171.

Estabrook, G. F. 1977. Does common equal primitive? *Syst. Bot.* 2:36–42.

Evans, H. E. 1973. Taxonomists' curiosity may help save the world. *Smithsonian* 4(6):36–43.

Exell, A. W. 1961. Malvaceae. In A. W. Exell and H. Wild, *Flora Zambesiaca* 1(2):420–511.

————, and D. Hillcoat. 1954. Um novo género de Malvaceae de Moçambique. *Estud. Ensay. Docum.* 12:55–61.

Eyde, R. H. 1971. Evolutionary morphology: Distinguishing ancestral structure from derived structure in flowering plants. *Taxon* 20:63–73.

————. 1976. The Seeds of Dicotyledons, a review. *Syst. Bot.* 1:195–196.

Fosberg, F. R. 1948. Derivation of the flora of the Hawaiian Islands. In E. C. Zimmerman, *Insects of Hawaii*, vol. 1, pp. 107–119. Honolulu: University of Hawaii Press.

————. 1972. The value of systematics in the environmental crisis. *Taxon* 21:631–634.

————, and M.-H. Sachet. 1966. *Lebronnecia*, gen. nov. (Malvaceae) des Iles Marquises. *Adansonia* n.s. 6:507–510.

————, and ————. 1972. *Thespesia populnea* (L.) Solander *ex* Correa and *Thespesia populneoides* (Roxburgh) Kosteletzky (Malvaceae). *Smithsonian Contr. Bot.* 7:1–13.

Fryxell, P. A. 1962. The "relict species" concept. *Acta Biotheor.* 15:105–118.

————. 1965a. A revision of the Australian species of *Gossypium* with observations on the occurrence of *Thespesia* in Australia (Malvaceae). *Austral. J. Bot.* 13:71–102.

————. 1965b. Stages in the evolution of *Gossypium. Advanc. Frontiers Plant Sci.* 10:31–56.

————. 1968a. A redefinition of the tribe Gossypieae. *Bot. Gaz.* 129:296–308.

————. 1968b. The typification and application of the Linnaean binomials in *Gossypium*. *Brittonia* 20:378–386.

————. 1969a. The West Indian species of *Gossypium* of von Rohr and Rafinesque. *Taxon* 18:400–414.

————. 1969b. A classification of the genus *Gossypium*. *Taxon* 18:585–591.

————. 1969c. The genus *Cienfuegosia* Cav. (Malvaceae). *Ann. Missouri Bot. Gard.* 56:179–250.

————. 1969d. The genus *Hampea* (Malvaceae). *Brittonia* 21:359–396.

————. 1971. Phenetic analysis and the phylogeny of the diploid species of *Gossypium* L. (Malvaceae). *Evolution* 25:554–562.

————. 1974. *Cienfuegosia* extended to Madagascar. *Ann. Missouri Bot. Gard.* 61:491–493.

————. 1976. *A nomenclator of* Gossypium—*the botanical names of cotton.* U.S. Dept. Agric. Tech. Bull. 1491. pp. 1–114.

————, and S. H. Hashmi. 1971. The segregation of *Radyera* from *Hibiscus* (Malvaceae). *Bot. Gaz.* 132:57–62.

————, and S. D. Koch. 1978. A range extension for *Gossypium lobatum* Gentry. *Southwestern Naturalist* 23:708–709.

————, and C. R. Parks. 1967. *Gossypium trilobum*: an addendum. *Madroño* 19:117–123.

————, and C. E. Smith, Jr. 1972. The contributions of Agostino Todaro to *Gossypium* nomenclature. *Taxon* 21:139–145.

Gaind, K. N., and S. C. Bapna. 1967. Antibacterial activity of *Thespesia populnea* Corr. *Indian J. Pharm.* 29:8–9.

Galichet, P. F. 1964. *Diparopsis watersi* Rotschild, Lepidoptera, Noctuidae, ravageur du cotonnier en Afrique Central. *Coton et Fibres Tropicales* 19:437–518.

Gammie, G. A. 1905. *The Indian cottons.* Calcutta: Government Printing Office. 38 pp. + 2 maps + 9 pls.

Garcke, A. 1860. Ueber die Gattung *Fugosia* Juss. *Bonplandia* 8:148–150.

Gray, A. 1855. Plantae novae Thurberianae. *Mem. Amer. Acad. Sci.* n.s. 5:297–328.

Hearn, A. B. 1968. Notes on *Gossypium* and *Cienfuegosia* in southern Yemen. *Cotton Growing Rev.* 45:287–295.

Hochreutiner, B. P. G. 1955. Malvacées. In H. Humbert, *Flore de Madagascar et des Comores.* Paris: Firmin-Didot. 170 pp.

Howard, R. A. 1949. *Atkinsia* gen. nov., *Thespesia*, and related West Indian genera of the Malvaceae. *Bull. Torrey Bot. Club* 76:89–100.

Hu, Shiu-ying. 1955. *Flora of China, Family 153 Malvaceae.* Arnold Arboretum. 80 pp. + 24 pls.

Hutchinson, J. 1967. *The Genera of Flowering Plants*, vol. 2. Oxford: Clarendon.

Hutchinson, J. B. 1943. A note on *Gossypium brevilanatum* Hochr. *Trop. Agric.* 20:4.

———. 1947a. Notes on the classification and distribution of genera related to *Gossypium*. *New Phytol.* 46:123–141.

———. 1947b. The classification of the genus *Gossypium*. In J. B. Hutchinson, R. A. Silow, and S. G. Stephens. *The Evolution of* Gossypium. London: Oxford University Press. xi + 160 pp.

Jacobs, M. 1966. On domatia—the viewpoints and some facts. *Proc. Koninkl. Nederl. Akad. Wetensch.* C69:275–316.

Johnson, B. L. 1975. *Gossypium palmeri* and a polyphyletic origin of the New World cottons. *Bull. Torrey Bot. Club.* 102:340–349.

Kearney, T. H. 1951. The American genera of Malvaceae. *Amer. Midl. Nat.* 46:93–131.

Knight, R. L. 1949. The distribution of wild species of *Gossypium* in the Sudan. *Empire Cotton Growing Rev.* 26:278–285.

Lewton, F. L. 1912. *Kokia*: A new genus of Hawaiian trees. *Smithsonian Misc. Coll.* 60(5):1–4 + 5 pls.

———. 1928. *Shantzia*, a new genus of African shrubs related to *Gossypium*. *J. Wash. Acad. Sci.* 18:10–16.

———. 1933. *Armouria*, a new genus of malvaceous trees from Haiti. *J. Wash. Acad. Sci.* 23:63–64.

Li, Hui-Lin. 1963. *Woody Flora of Taiwan*. Morris Arboretum.

Martin, A. C. 1946. The comparative internal morphology of seeds. *Amer. Midl. Nat.* 36:513–660.

Mauer, F. M. 1954. *Proiskhozhdeniye i Sistematiki Khlopchatnika* [Origin and Systematics of Cotton]. Tashkent: Akad. Nauk Uzbek. S.S.R. 384 pp. [In Russian.]

Mayers, W. F. 1868. Cotton in China. *Notes & Queries on China & Japan* 2:72–74, 94–95.

McVaugh, R. 1945. The genus *Triodanis* Rafinesque and its relationships to *Specularia* and *Campanula*. *Wrightia* 1:13–52.

Menninger, E. A. 1962. *Flowering Trees of the World*. New York: Hearthside Press. 336 pp.

Muramoto, H. 1969. Hexaploid cotton: Some plant and fiber properties. *Crop Sci.* 9:27–29.

Parkin, J. 1953. The Durian Theory—a criticism. *Phytomorphology* 3:80–88.

Parlatore, F. 1866. *Le Specie dei Cotoni*. Firenze: Stamperia Reale. 164 pp.

Phillips, L. L. 1963. The cytogenetics of *Gossypium* and the origin of New World cottons. *Evolution* 17:460–469.

Pickersgill, B., S. C. H. Barrett, and D. de Andrade-Lima. 1975. Wild cotton in northeast Brazil. *Biotropica* 7:42–54.

Pijl, L. van der. 1966. Ecological aspects of fruit evolution: A functional study of dispersal organs. *Proc. Koninkl. Nederl. Akad. Wetensch.* C69:597–640.

———. 1969. *Principles of Dispersal in Higher Plants*. Berlin: Springer-Verlag. 154 pp.

Proctor, V. W. 1968. Long-distance dispersal of seeds by retention in digestive tract of birds. *Science* 160:321–322.

Prokhanov, J. I. 1947. The conspectus of a new system of cotton (*Gossypium*). *Bot. Zhurn.* 32:61–78. [In Russian and Latin with an English summary.]

Raven, P. H., B. Berlin, and D. E. Breedlove. 1971. The origins of taxonomy. *Science* 174:1210–1213.

Record, S. J. 1935. Note on the wood of *Cephalohibiscus*. *Trop. Woods.* 44:21.

Reeves, R. G. 1936. Comparative anatomy of the seeds of cottons and other malvaceous plants II: Hibisceae. *Amer. J. Bot.* 23:394–405.

Roberty, G. 1942, 1946, 1950. Gossypiorum revisionis tentamen. *Candollea* 9:19–103; 10:345–398; 13:9–165.

Rohr, J. B. P. von. 1791–1793. *Anmerkungen über den Cattunbau, zum Nuzen der Dänischen Westindischen Colonien.* Altona and Leipzig: Hammerich.

Sauer, J. 1967. *Geographic Reconnaissance of Seashore Vegetation along the Mexican Gulf Coast.* Coastal Studies Institute, Louisiana State University, Tech. Report no. 56. x + 59 pp.

Saunders, J. H. 1961. *The Wild Species of* Gossypium *and Their Evolutionary History.* London: Oxford University Press.

Schumann, K., and M. Gürke. 1891–1892. Malvaceae I & II. In C. F. P. von Martius, *Flora Brasiliensis* 12(3):253–624.

Smith, A. C. 1969. Systematics and appreciation of reality. *Taxon* 18:5–13.

Sporne, K. R. 1954. Statistics and the evolution of dicotyledons. *Evolution* 8:55–64.

Standley, P. C. 1927. The genus *Hampea*. *J. Wash. Acad. Sci.* 17:394–398.

Stebbins, G. L., Jr. 1947. Evidence on rates of evolution from the distribution of existing and fossil plant species. *Ecol. Monogr.* 17:149–158.

_____. 1959. The role of hybridization in evolution. *Proc. Amer. Philos. Soc.* 103:231–251.

_____. 1974. *Flowering Plants: Evolution above the Species Level.* Cambridge, Mass.: Belknap Press of Harvard University Press. xviii + 399 pp.

Stuessy, T. F. 1975. The importance of revisionary studies in plant systematics. *Sida* 6:104–113.

Swartz, O. 1790. Botaniske Anmärkningar om Bomullsslagen. *Kongl. Vetensk. Acad., Nya Handl.* 11:20–25.

Takhtajan, A. 1969. *Flowering Plants: Origin and Dispersal.* Trans. from Russian by C. Jeffrey. Washington, D.C.: Smithsonian Inst. Press. x + 310 pp.

Ter-Avanesyan, D. V. 1973. *Khlopchatnik* [Cotton]. Leningrad: Kolos. 483 pp.

Todaro, A. 1863–1864. Osservazioni su talune specie di Cotone coltivate nel R. Orto Botanico di Palmero. *Giorn. R. Ist. Incorr. Agric. Art. Manif. Sicil.* ser. 3, 1:1–15, 33–120.

_____. 1877. *Relazione sulla cultura dei Cotoni in Italia seguita da una Monografia del genere* Gossypium. Rome: Molina. iii + 287 pp.

_____. 1878. Tavole (plates to accompany the preceding volume). Palermo: Visconti. 9 pp. + 12 pls.

Torrey, J. 1859. *Botany of the United States and Mexican Boundary Survey.* Washington, D.C.: Government Printing Office. 270 pp. + 61 pl.

Ulbrich, E. 1914. Über einige Malvaceen-Gattungen aus der Verwandtschaft von *Gossypium* L. *Bot. Jahrb.* 50 (suppl.):357–362.

Uphof, J. C. T. 1938. Cleistogamic flowers. *Bot. Rev.* 4:21–49.

Urban, I. 1912. *Maga. Symbol. Antill.* 7:281–282.

————. 1924. Plants from Beata Island, Santo Domingo, Phanerograms. *Dansk Bot. Arkiv* 4(7):5–10 + 3 pls.

Valíček, P. 1974. *Plané a Kulturní Bavlníky* [Wild and Cultivated Cottons]. Prague: Inst. Trop. Subtrop. Zemědělství. 206 pp.

Varuntsyan, J. S. 1958. Nekotoriye problemi klassifikatsii khlopchatnika [Some problems of classification of cotton]. *Agrobiology* 5(113):141–156. [In Russian; translation available by C. C. Nikiforoff.]

Vavilov, N. I. 1929. Gavriil Semenovich Zaitzev, 1887–1929. *Bull. Appl. Bot. Genet. Plant Breeding* 2(5):iii–xvi. [In Russian with bibliography of Zaitzev's publications.]

Vichkanova, S. A., and L. Y. Goryunova. 1968. Izuchenie virulitsidnoi aktivnosti gossipola in vitro [In-vitro study of the viricidal activity of gossypol]. *Antibiotiki* 13:828–829. [In Russian.]

Vijendra Das, L. D., and S. W. Mensikai. 1968. Pollen mother cell analysis of intraspecies hybrid of *Gossypium arboreum* L. indicating secondarily balanced polyploidy. *Cytologia* 33:188–194.

Walters, S. M. 1961. The shaping of angiosperm taxonomy. *New Phytol.* 60:74–84.

Watt, G. 1907. *The Wild and Cultivated Cotton Plants of the World.* London: Longmans, Green & Co. x + 406 pp.

Wight, R. 1838–1853. *Icones Plantarum Indiae Orientalis.* Madras, India. 6 vols.

Wilson, F. D., J. A. Lee, and R. R. Bridge. 1968. *Genetics and Cytology of Cotton 1956–67.* Southern Coop. Ser. Bull. 139. 84 pp.

Zaitzev, G. S. 1928. A contribution to the classification of the genus *Gossypium* L. *Bull. Appl. Bot. Genet. Plant Breeding* 18:39–65. [In Russian on pp. 1–38, with illustrations.]

Index

Page references in italics are to the principal taxonomic references; those with an asterisk (*) indicate illustrations.

Abelmoschus zollingeri, 99
acarids (mites), 112
acarodomatia, 112
Acaulimalva, 186
adaptation, xvii, 107, 114, 133–138, 140–144, 148, 150–153, 157–158, 186–190, 196, 199, 205–207
Aden, 27
advanced characters, 136, 185, 193, 208–213
Adventives, 160
Africa, 9, 16, 26–28, 37, 64–68, 89, 93, 99, 130, 132, 166–168, 172, 187–189, 194, 198, 202, 205
agriculture, origin of, 164, 168, 173–174, 176
Alefeld, F. G. C., 3–4
Aliotta, A., 6, 13
Althaea, 3
Alyogyne, 14
Amazon Basin, 176
amphidiploid, 11, 145–146, 163–164, 170–171, 181, 183, 201
Anderson, E., 3, 17, 159
androdioecy, 121, 215–216
androecium, 121–122, 192, 196–197, 199–200, 210, 220
aneuploid series, 180
angiosperms, origin of, 186
Angola, 28, 65, 144, 189
animals, 136–139, 143, 168
annual habit, 151–152, 169–170
Anotea flavida, 199
anthers, 121, 122*, 194, 197
anthocyanins, 204
Anthonomus grandis, 160, 173–174, 205
Antilles, 71. *See also* Caribbean, Cuba, Puerto Rico, Danish West Indies, St. Croix, St. Thomas, Dominican Republic
Arabia, 27, 65–68, 130, 160, 167, 188–191, 198
archaeological evidence, 164, 167–168
Argentina, 32–33, 35, 142

aridity, 113, 143–145, 151, 155, 161, 187–191, 200
aril, 127, 134–139
Aristolochos of Paros, 178
Arizona, 56, 150–152, 155, 186
Armouria, 84
    beata, 89
Aruba Island, 143
Atkinsia, 15, 84, 100
    cubensis, 89
Atlantic Ocean, 143
Australasia, 131, 132
Australia, 50, 52–53, 99, 111, 132, 144, 147, 151, 155–156, 160, 162, 186, 189
Azanza, 15, 84
    acuminata, 99
    garckeana, 93
    lampas, 95
    zollingeri, 99

Bahama Islands, 28, 143, 188
Baja California, 56–58, 144, 191
Baker, E. G., 14, 100
Baker's Law, 148
bark, as source of dye, 159, 203
    as source of fiber, 161
Batesimalva, 108
Beasley, J. O., 11*–12, 163
Beasley Laboratory, Texas A&M University, 12
Beata Island (Isla Beata), 93, 196, 199
Beckett, R. E., 218
bees, 195
Bell, A. A., 207
Bentham, G., 4, 100
Bering land bridge, 153
birds, 135, 137, 139, 142–143
Bolivia, 30, 35
boll weevil. *See Anthonomus grandis*
Bombacaceae, 15, 115, 123, 135, 136, 162
Bombax, 123
Borneo, 99
Borssum Waalkes, J. van, 15, 16, 100, 211

Botswana, 166
Bougainville, 132
bracts, of the involucel, 115*, 116–118,
    216–217
Brazil, 30, 32, 71, 164, 170, 199
breeding. *See* selection
bridging species, 182
British Honduras, 78
Bukasov, S. M., 12
Bupariti, 14, 84
Burma, 169

Callirhoë, 121
calyx, 115*, 118–119, 210, 213, 215
Cape Verde Islands, 65, 144, 189
capsules, 100–101. *See also* fruits
Capuron, R., 100
Caribbean, 29, 70–71, 140, 144, 146, 160,
    162, 171, 196
Cavanilles, J., 14
Central America, 9, 71, 75, 78–79, 132,
    135, 144, 162. *See also* Middle
    America
Cephalohibiscus, 16, 19, 106, 113, 117,
    119, 131–132, 157, 179, 187, 192,
    195–196, 211, 213–214, 220
    peekelii, 20, 21*, 99, 104, 107, 109, 117,
        122, 124–127, 133, 141, 162,
        180, 197, 199–200
Charadriiformes, 142
Chiapas, 75, 78
Chihuahua, 56
China, 169–170
Chromosome homology, 162, 181–183
    morphology, 181–184
    numbers, 16, 18–19, 22, 36, 40–41, 46,
        72, 80, 84, 136, 163, 178–180,
        182, 211, 219
Cienfuegia, 21
Cienfuegosia, 14, 18–19, 21, 22 (key to
    spp.), 130, 131 (map), 132, 151,
    178–179, 181, 183–184, 187,
    192–193, 196, 203, 205, 211,
    213–214, 219
    subgen. *Articulata*, 22, 26, 180, 193,
        202
    subgen. *Cienfuegosia*, 22, 28, 109, 111,
        113, 193, 202
    sect. *Articulata*, 23, 26, 113, 132, 191
    sect. *Cienfuegosia*, 24, 28, 118, 125,
        187–188, 191, 193, 202, 223

sect. *Dioica*, 23, 27
sect. *Friesia*, 25, 35, 110, 118, 124, 185,
    187, 193, 197
sect. *Garckea*, 23, 26, 110, 127, 188
sect. *Paraguayana*, 25, 32, 124, 187,
    193, 197
sect. *Robusta*, 25, 30, 116, 124–125,
    193
sect. *Spathulata*, 25, 32
affinis, 30, 31, 108, 109*, 122*, 123,
    125–126, 133, 139, 156, 160,
    179, 187, 195, 205, 220
    var. *campestris*, 30
    var. *humilis*, 30
argentina, 35, 106, 108, 124, 127, 179,
    194
    var. *hasslerana*, 35
australis, 53
benthamii, 53
bricchettii, 65
chiarugii, 27
cuyabensis, 30
digitata, 14, 22, 28, 109*, 110, 122, 143,
    156, 179, 188
    var. *lineariloba*, 28
drummondii, 32, 34*, 109*, 115*,
    122–123, 127, 133, 141–142,
    179, 220
    var. *genuina*, 32
    var. *pubescens*, 32
ellenbeckii, 65
flaviflora, 99
gerrardii, 26, 99, 108, 110, 125, 180,
    185, 187, 217
glabrifolia, 30
gossypioides, 48
hasslerana, 35, 110
hearnii, 27*, 104*, 119, 126, 179, 220
heteroclada, 27, 28*, 114, 117, 120,
    141, 187, 194–195
heterophylla, 28, 29*, 110, 116, 124,
    127, 143, 179, 188, 194
    ssp. *subternata*, 29
    var. *cuneata*, 30
hildebrandtii, 26, 109*, 110, 116–117,
    122*, 133, 141, 179, 185
hispida, 35
hitchcockii, 32, 108, 180, 192
humbertiana, 35, 110, 180
incana, 65
integrifolia, 33, 187

intermedia, 32
junciformis, 28
    var. *ruyssenii*, 28
latifolia, 52
pedata, 53
pentaphylla, 64
phlomidifolia, 30
    var. *humilis*, 30
populifolia, 52
pulchella, 52
punctata, 53
riedelii, 30
robinsonii, 52
rosei, 28, 108, 110, 121, 127, 156, 179, 195, 205
somalensis, 65
somaliana, 27, 108, 123, 179
subprostrata, 33, 125
subternata, 29, 110, 188, 193
sulfurea, 32, 33*, 107–108, 125, 179, 194
sulphurea, 30, 32
    var. *drummondii*, 32
    var. *genuina*, 32
    var. *glabra*, 32
    var. *integrifolia*, 33
    var. *major*, 32
thespesioides, 99
tripartita, 30, 108, 110, 179, 188, 193
triphylla, 53
ulmifolia, 35*, 115*, 179
welshii, 26, 125, 179
yucatanensis, 28, 108–109, 115*, 119–120, 129, 143 (map), 179, 187–188, 194, 197, 220
Citrus, 171
classification, xiii–xvi, 12–13, 71, 166, 222, 224–225
    naturalness of, xv, 4, 13, 220
cleistogamy, 197, 200
Clementsian succession, 148
coevolution, xvii, 138, 190, 206, 221
Colima, 78, 113
Colombia, 79, 143
Constance, L., xvii
convergent evolution, 209
Corner, E. J. H., 135–138
corolla, 119–120, 198, 210
    funnelform, 119, 198, 220
    rotate, 119, 194, 199, 220
    tubular, 199, 221

Costa Rica, 78–79, 135
cotton gin, 173–175
cotton leafworm, 206
cottonseed oil, 106, 173, 175, 203, 205
cotyledons, 103, 106–107, 210
crepuscular pollinators, 195–199
Cretaceous, 132, 163, 202
Cronquist, A., 134, 136–137
Crowson, R. A., 219
crystal layer, of seed coat, 128–129
Cuba, 28, 89, 143, 188
cultural diffusion, 168
Curaçao, 145
cuttings, of Gossypieae, 196
cytogenetics, 9, 11, 162, 181–182
cytology (karyology), 9, 178, 180
cytotaxonomy, 178, 180

Danish West Indies, 5
darkling beetles, 205
Darwin, C., xiii
Davis, P. H., 222
day length, 151, 154, 156–158, 169–170
Decaisne, J., 17
DeCandolle, A., 3, 103
Decaschistia, 111
Degener, O., 15–16
descriptive biology, xvii–xviii
deserts, 139, 141, 151, 188–190, 196
dioecism, 114, 120–121, 123, 194–196, 200, 215–216
Diparopsis, 205
Diredawa, Ethiopia, 27
diseases. *See* pests and diseases
dispersal, of fruit, 118, 125, 133, 140–141, 147
    of seed, 125, 133, 135–142, 146–147, 221
    by man, 132, 145, 162, 167
    by sea, 132, 140, 143–144, 147–148, 165, 187–188, 201–202
    mechanisms of, xvii, 142, 145
Dobzhansky, Th., 185
domatia, 102, 112*, 113
domestication, of cotton, 164–170, 173
Dominican Republic, 93
dormancy, of plants, 150–153, 156, 186
    of seeds, 104, 127–128, 139, 141, 166, 169
Drosophila, xvii
Duhamel, H. L., 14

Dumont, A., 4
duplication, genetic, 183
durian theory, 135–136
Durio zibethinus, 135–137
dye, 159, 203

East Africa, 16, 26, 37, 68, 89, 93, 99, 189
ecological adaptation, 148, 186–190
ecologists, 224
ecology, xvi–xviii, 9, 103, 139, 144–145,
    147, 164, 170, 196–197, 201, 223
Ecuador, 30, 32, 188
elevation (above sea level), 150–151,
    186–187, 190
Elidurandia, 21
    texana, 32
El Salvador, 78
embryo, 3–4, 103, 104*, 105*, 128–129,
    147, 206–207, 210
endosperm, 104, 106, 210
Endress, P. K., 136
Eocene, 153
epicotyl, 104, 106
epigeal germination, 210
Erioxylum, 39
    aridum, 58
Ethiopia, 27, 167
evolution, xvi–xvii, 103, 146, 152, 158,
    169, 176–177, 208, 225
    convergent, 209
    parallel, 137, 168, 198, 219–221
    rates of, 164, 185, 200–202
    reticulate, 209
evolutionary canalization, 133, 153, 158
    continuum of, 221–223
    history of, xvii–xviii, 131, 168, 175–178,
        209, 220
    opportunism in, 144, 146–147
    plasticity of, 220
    rates of, 164, 185, 200–202, 208
    "tree" of, 222
    trends in, 185–186, 188, 191–194, 198,
        219
Exell, A. W., 15, 100
extinction, 16, 80–82, 132, 134, 159–160
Eyde, R. H., 103, 208

Far East, 64, 169
Ferguson Island, 140

fiber, color of, 126
    quality of, 173
    See also seed hairs
filaments, 121, 210
floral biology, 155, 190–210
floral initiation, 151, 153–157, 189
Florida, 28, 143, 147, 162, 171, 188, 196
flower buds, 115*, 216
flowers, 118–124, 122*, 155, 210, 213
    of dioecious species, 120*
    odor of, 120, 190, 195, 197
Fosberg, F. R., 14, 100
fossils, 221
    absence of, 131, 164, 185, 200, 202,
        208–209
fragrance, floral, 120, 190, 195, 197
Frankie, G., 135
frost, 131, 150–153, 155, 157, 169, 186
fruits, 100–101, 124–126, 135–136,
    138–142, 147, 150, 152, 155,
    169–175, 196–197, 210, 215
Fugosia, 14, 21
    affinis, 30
    areysiana, 67
    argentina, 35
    australis, 53
    campestris, 30
    cuneata, 30
    digitata, 28
    drummondii, 32
    flaviflora, 99
    gerrardii, 26
    guianensis, 30
    heteroclada, 27
    heterophylla, 28
    lanceolata, 30
    latifolia, 52
    pedata, 53
    phlomidifolia, 30
    populifolia, 52
    pulchella, 52
    pulverulenta, 32
    punctata, 29, 52
    retusa, 30
    sulfurea, 32
        var. *trifida*, 32
    thespesioides, 99
    tripartita, 30
    triphylla, 53
    welshii, 26

Galápagos Islands, 56, 71, 143–144, 160, 164, 170
Garcke, A., 14
geneticists, 163, 224–225
genetics, xvi, 9, 120, 134, 142, 161, 166, 169, 193–195, 204, 206–207, 223, 225
genome groups, 162, 181–182
germination, 104, 106, 127–128, 135, 141–143, 147, 164, 169, 176, 210
Ghana, 27
Gilia, xvii
glaciation, 145, 201
glandless cotton, 204–207
Gondwanaland, 132, 201–202
Goodenough Island, 140
Gossypieae, origins of, 131–132, 158
Gossypioides, 16, 19, 36 (incl. key to spp.), 108–110, 113, 117, 119, 130, 178, 183, 185, 187, 191, 205, 211, 213–214
  brevilanatum, 37, 39*, 142, 179, 183, 218
  kirkii, 36, 37, 38*, 110–111, 126, 139, 179, 218
Gossypium, xvii, 3, 5–14, 19, 37, 41 (key to spp.), 105*, 113, 116–117, 130–132 (map), 162, 164 (map), 178, 180–182, 185, 192, 196, 200–202, 205, 211–214, 223
  subgen. *Gossypium*, 46, 62
  subgen. *Houzingenia*, 43, 53, 152
  subgen. *Karpas*, 46, 68
  subgen. *Sturtia*, 41, 48
  sect. *Erioxylum*, 43, 58, 107, 113, 212
  sect. *Gossypium*, 46, 62
  sect. *Grandicalyx*, 42, 52, 191
  sect. *Hibiscoidea*, 42, 53
  sect. *Houzingenia*, 44, 53
  sect. *Pseudopambak*, 46, 65
  sect. *Sturtia*, 41, 48
  subsect. *Anomala*, 46, 64
  subsect. *Austroamericana*, 45, 62
  subsect. *Caducibracteolata*, 43, 56, 189, 191
  subsect. *Erioxylum*, 45, 58, 117, 150, 155, 189, 191
  subsect. *Gossypium*, 46, 62
  subsect. *Houzingenia*, 44, 53
  subsect. *Integrifolia*, 44, 56

  subsect. *Longiloba*, 47, 68
  subsect. *Pseudopambak*, 47, 65, 191
  subsect. *Selera*, 45, 58
  abyssinicum, 64
  acuminatum, 70
  africanum, 64
  albiflorum, 62
  album, 64
  anomalum, 64, 117, 144, 179
    ssp. *areysianum*, 67
    ssp. *steudneri*, 64
    ssp. *triphyllum*, 53
  arboreum, 40, 62, 116, 153, 166, 168–170, 218
  areysianum, 67*, 119, 125, 179
  aridum, 58, 59*, 119, 126, 155, 179, 189, 212, 218, 220
    var. *palmeri*, 58
  armourianum, 57, 117–118, 126–127, 141, 179, 189, 198
  asiaticum, 68
  australe, 53, 111, 115*, 118, 125, 127–128, 133, 141, 156, 197, 205
  australiense, 48
  bani, 64
  barbadense, 13, 70, 112, 123, 143, 145, 161, 164, 168–171, 176, 179, 218
    ssp. *darwinii*, 71
    var. *braziliense*, 176
    var. *darwinii*, 71, 170
  barbosanum, 65
  benadirense, 65
  bickii, 53, 111, 117, 127, 153, 156, 179, 197, 205
  birkinshawii, 70
  brasiliense, 70
  brevilanatum, 37
  bussei, 37
  caespitosum, 70
  caicoense, 71
  californicum, 57
  capitis-viridis, 65, 110, 117, 144, 179, 189
  cernuum, 62
  convexum, 70
  costulatum, 52, 139
  croceum, 62
  cunninghamii, 52, 109*, 126–127, 144, 189

darwinii, 71, 143, 145, 160, 164, 171, 179
davidsonii, 56, 109, 117, 126, 139, 144, 179, 190
divaricatum, 70
drynarioides, 80, 82
ellenbeckii, 65
flaviflorum, 99
frutescens, 64
fruticulosum, 68
gossypioides, 48, 58, 61*, 109, 111, 117–118, 122, 125, 179, 190
guyanense, 70
harknessii, 56, 118, 124–127, 133, 141, 179, 189, 220
    ssp. *armourianum*, 58
harrissii, 70
herbaceum, 64, 153, 166–170, 179
    var. *africanum*, 166, 168, 171
    var. *steudneri*, 64
hirsutum, 68, 115*, 126, 143–148, 161, 164, 168–173, 179, 204–205, 218
    var. *palmeri*, 171
    forma *tomentosum*, 68
hopi, 70
incanum, 65, 119
indicum, 62
intermedium, 62
jamaicense, 68
janiphaefolium, 68
kirkii, 37
    ssp. *brevilanatum*, 37
    ssp. *scandens*, 37
klotzschianum, 56, 126, 139, 144, 179
    ssp. *raimondii*, 62
    var. *davidsonii*, 56
lanceiforme, 55
lanceolatum, 68, 71, 164, 171, 179
lapideum, 70
latifolium, 68
laxum, 58, 108–109, 155, 189–190, 198, 212
lobatum, 58, 60*, 108, 114, 115*, 118–119, 126–128, 155, 179, 189, 191, 212
longicalyx, 68, 69*, 108, 117–118, 189, 197–198
marie-galante, 70
maritimum, 70
mexicanum, 70
micranthum, 68

microcarpum, 64
multiglandulosum, 70
mustelinum, 71, 164, 170–171, 179
nandewarense, 50
nanking, 62
neglectum, 62
nelsonii, 53, 127, 141, 197
nervosum, 70
nicaraguense, 70
nigrum, 62
obtusifolium, 62
oligospermum, 68
pallens, 68
paniculatum, 68, 179
paolii, 65
pedatum, 70
perenne, 70
perrieri, 64
peruvianum, 70
pilosum, 52, 107
populifolium, 52, 107–108, 126–127, 139, 144, 185
prostratum, 68
pulchellum, 52, 189
punctatum, 68
puniceum, 62
quinacre, 70
raimondii, 62, 63*, 117, 127, 179, 190
religiosum, 68
robinsonii, 50, 51*, 106, 109*, 110–111
rohrianum, 70
rosei, 58
roxburghii, 64
royleanum, 62
rubrum, 62
rufum, 62
sandvicense, 68
sanguineum, 62
schottii, 70
senarense, 64
sericatum, 70
siamense, 68
somalense, 65, 109*, 118, 125, 179, 197–198, 210
soudanense, 64
speciosum, 62
stocksii, 65, 66*, 108, 110, 123, 179, 185, 189
sturtianum, 48, 49, 106, 108, 117–118, 129, 151–152, 155, 162, 169, 179, 186, 218

ssp. *robinsonii*, 52
var. *nandewarense, 50**, 110, 117, 189
sturtii, 48
ssp. *robinsonii*, 52
suffruticosum, 70
taitense, 70
thespesioides, 99
thurberi, 56, 57*, 100, 108, 109*, 110–111, 115*, 118, 120, 126, 128, 133, 139, 150–155, 169, 179, 186, 190, 198, 204–205
tomentosum, 68, 109, 111, 124, 145, 154, 160–161, 164, 170–171, 179, 196, 198, 220
var. *parvifolium*, 68
transvaalense, 64
trilobum, 55*, 56, 62, 108, 110, 119, 124, 126, 139, 152, 179, 186, 198, 220
triphyllum, 53, 54*, 108, 109*, 110, 125, 127, 179
turneri, 58, 118, 141, 189
vaupellii, 70
vitifolium, 70
volubile, 70
walchottianum, 50
wattianum, 64
wightianum, 64
zaitzevii, 64
gossypol, 159, 203–207
glands, 3–4, 118, 204, 206–207
medicinal uses of, 203
graft compatibilities, 218
growth habit, 107–108, 114, 139, 169–170, 173–175, 210
evolutionary trends in, 185, 212
See also annual habit
Guatemala, 75–79
Guaymas, 58
Guerrero, 58
Gulf of Mexico, 143, 144, 146, 171, 187
gynodioecism, 121, 195, 200

Hampea, 15, 19, 72 (incl. key to spp.), 107, 114, 120–123, 127, 130–131, 134, 138, 156, 161, 178, 180, 187, 191, 195–196, 198, 211–214, 219, 223
sect. *Hampea*, 73, 75

sect. *Standleya*, 73, 78, 121, 192, 195, 213, 215
sect. *Trianchonia*, 72, 78, 110
series *Hampea*, 75
series *Preslia*, 75
series *Watsonia*, 78
albipetala, 79, 107
appendiculata, 79, 104*, 107, 135
var. *longicalyx*, 79
bracteolata, 75
breedlovei, 75, 114, 138, 210
dukei, 79
euryphylla, 78
integerrima, 72, 75, 109, 112, 115*, 179
var. *appendiculata*, 79
latifolia, 78, 112, 114, 210, 213, 217
longipes, 75, 76*, 109*, 112, 114, 186
macrocarpa, 78
mexicana, 78, 109*, 187
micrantha, 78, 109, 114, 192, 215
montebellensis, 75
nutricia, 75, 77*, 107, 120*, 123, 125, 135, 179, 205, 213
ovatifolia, 79
panamensis, 79
platanifolia, 78, 109, 111, 119, 125, 127, 213
punctulata, 79
romeroi, 79
rovirosae, 78, 107, 114–116, 122, 124, 179, 213, 215, 217
sphaerocarpa, 78, 118, 125
stipitata, 78, 118, 179
thespesioides, 79, 107, 186
tomentosa, 78, 100, 111, 112*, 113, 116–117, 121, 125, 127, 179, 187, 215–217
trilobata, 78, 109*, 111, 126
hard-seededness, 127, 129. *See also* seed coat, impermeability of
Hawaii, 15, 68, 80–82, 154, 159–162, 164, 170, 203
Heliothis, 205
hexaploids, 152, 182
Heywood, V. H., 222
Hibisceae, 3–4, 104, 133
Hibiscus, 3, 104, 111, 194
sect. *Furcaria*, 115
sect. *Lampas*, 95
affinis, 30

anomalus, 64
archboldianus, 107
argentinus, 35
australis, 53
bacciferus, 86
barbadensis, 70
blumei, 87
caesius, 104*
callosus, 95
calyphyllus, 115
campestris, 30
campylosiphon, 99
cannabinus, 104*
cavanillesii, 28
cernuus, 64
cuneatus, 30
drummondii, 32
drynarioides, 82
flaviflorus, 99
fruticulosus, 70
gangeticus, 95
glabrifolius, 30
gossypioides, 48
hilairei, 30
hitchcockii, 32
humbertianus, 36
ingenhoussii, 55–56
jussieui, 32
lampas, 95
latifolius, 52
meraukensis, 104*
nanking, 64
peekelii, 20
phlomidifolius, 30
poeppigii, 199, 221
populifolius, 52, 86
populneoides, 89
populneus, 14, 86
preslii, 78
pulverulentus, 32
rectiflorus, 30
redoutei, 28
religiosus, 70
robinsonii, 50
sulphureus, 30
     var. *acutifolius*, 30
tetralocularis, 95
tiliaceus, 107
tomentosus, 68
tripartita, 30
welshii, 26

Hill, S. R., xiii
Hillcoat, D., 15, 100
Honduras, 75, 78
Hooker, J. D., 4, 100
host-plant resistance, 173
Howard, R. A., 15, 100
humans
   curiosity of, 166
   influence of, xviii, 145–147, 159, 177,
      186
   population pressures of, xviii, 176
hummingbirds, 199–200, 221
Hutchinson, J., 100
Hutchinson, J. B. (Sir Joseph), 4, 9, 10*,
      11–14, 100, 183, 218
Huxley, J., 150
hybrids, 9, 161–163, 181–183, 202, 218
Hymenoptera, 195, 199
hypocotyl, 104, 106
hypogyny, 210

identification, xiii–xiv
India, 7, 64, 99, 147, 162, 167–169
Indian Ocean, 89, 148
Indochina, 9, 99
Indonesia, 169
industrialization, of cotton agriculture,
      173–176
inflorescence, 113–114, 190, 192, 198,
      213, 215
Ingenhouzia, 39
   harknessii, 56
   triloba, 55
introgression, 161
inversions, 182
involucel, 101, 114, 118, 198, 216–217.
      See also bracts, of the involucel
Isla Beata, 93
isolation, 16, 201
Italy, 5–6

Jacobs, M., 112–113
jassids, 173
Java, 99
Johnson, B. L., 171
Jussieu, A. L. de, 14

kapok, 162
karyology. *See* cytology
karyotype analysis, 181, 183–184
Kauai, 82, 159

Kearnemalvastrum, 186
Kearney, T. H., 12, 100
Kenya, 37, 65
keys, xv
  to genera, 18–19
  to species of *Cienfuegosia*, 22–26
  to species of *Gossypioides*, 36–37
  to species of *Gossypium*, 41–48
  to species of *Hampea*, 72–75
  to species of *Kokia*, 80
  to species of *Thespesia*, 84–86
kidney cotton, 175–176
Kokia, 15–16, 18, 79, 80 (key to spp.), 107,
    109, 113, 117, 119, 125–126,
    131–132, 142, 157, 178, 183, 187,
    192, 199, 203, 211–214
  cookei, *81*, 109*, 159, 179
  drynarioides, *81**, 118, 159, 162, 179
    var. *lanceolata*, 80
  kauaiensis, *82**, 118, 159, 220
  lanceolata, *80*, 159
  rockii, 82, 218
    var. *kauaiensis*, 82
Komarov Botanical Institute, 12
Korea, 169–170
Kruckeberg, A. R., 219
Kydia, 120

leaf margins, 110
leaves, 101, 107–108, 109*, 110–113,
    151–152, 155, 188–189, 210
Lebronnecia, 16, 19, *83*, 107, 113, 117,
    119, 131–132, 138, 157, 178, 187,
    191, 195, 210–214, 217
  kokioides, *83*, 109*, 111, 115, 123, 125,
    127, 129, 142, 160, 180
Leguminosae, 135, 218
Lepidoptera, 195, 197, 199, 205
Levant, 167
Lewton, F. L., 15, 100
lignin, 127–129
Linnaeus, C., xvii, 5
lint percentage, 164, 166, 172–175
littoral habitats, 113, 142–149, 164,
    171–172, 175, 188, 201
longevity, of seeds, 128
Luehea divaricata, 99
lygus bugs, 205

Madagascar, 16, 36–37, 93

Maga, 84
  cubensis, 89
  grandiflora, 93
maize, xv
majagua, 161
male sterility, 121, 195
Mali, 188–189
Malvaviscus, 199, 221
  populneus, 86
Malveae, 111, 120–122
mammals, 135, 137
Mangelsdorf, P., xv
Margarita Island, 143
Marquesas Islands, 16, 70, 83, 143,
    160–161
Martin, A. C., 103
maternal inheritance, 195
Matto Grosso, 32
Mauer, F. M., 10–13
Mauritania, 188
McMichael, S., 204
mechanical harvesting, 173–174
Melanesian Islands, 141
Mexico, 28, 32, 56–62, 68, 75–78, 107,
    113, 144, 146, 152, 171, 188, 191,
    199
Michoacán, 58
Middle America, 70, 130, 161, 164, 171.
    See also Central America
Middle East, 64, 167, 169
mites (acarids), 112
Mohenjo Daro, 167
Molokai, 81
monoculture, 173
Montezuma, 15, 84, 100
  cubensis, 89
  grandiflora, 93
  speciosissima, 93
Morelos, 56
moths, 195, 197, 199
Mozambique, 89, 166
Muramoto, H., 152

Namibia, 65, 144, 189
Napaea dioica, 120
Natal, 37, 89
natural history, xvi–xviii, 223–225
natural selection, 161, 166. *See also*
    adaptation, evolution

nectaries, of the calyx, 115, 190
  extrafloral, 210
  foliar, 108, 111–112
  involucellar, 111, 115–116, 213,
    215–217
  stipular, 111
Neesia altissima, 99
Neogossypium, 40
New Guinea, 15, 21, 89, 93, 107,
    131–132, 140, 160, 162, 196
New Ireland, 132
New World cottons, 9, 11, 163, 166–171,
    181, 201
Nicaragua, 78
Nigeria, 27–28, 65
nomenclature, xiv, 5, 14–15, 181
Normanby Island, 140
Nototriche, 186
Notoxylinon, 40
  australe, 53
  flaviflorum, 99
  latifolium, 52
  pedatum, 53
  punctatum, 53
  robinsonii, 52
  thespesioides, 99
Nubia, 168

Oahu, 80
Oaxaca, 28, 58, 62, 78, 188
odor, floral, 120, 190, 195, 197
Oenothera, xvii
Okinawa, 169
Old World cottons, 9, 163, 166, 168–169,
    181, 201
Oman, 65
ornamentals, 118, 140, 159, 162, 190, 199
outbreeding, 124, 194, 198, 200, 221
outpost shrubs, 148, 171
ovary, 124

Pachira, 123
Pacific Ocean, 70, 143–144, 146, 161
Pakistan, 65, 189
palaeoclimate, 131
Palermo Botanic Garden, 5
palisade, of leaf mesophyll, 189
  of seed coat, 128
palynology, 123
Panama, 78–79
Papua, 140

Paraguay, 30, 32–33, 35, 142, 188
Paritium gangeticum, 95
Parkin, J., 135
Parlatore, F., 3, 5–6, 13
Pavonia, 3
  macrostyla, 199
pedicel, 114–118
pentaploids, 182
Periptera, 199, 221
Peru, 30, 32, 62, 188
pesticides, 173, 206
pests and diseases, 160, 173–174, 205–207
petals, 119, 190, 192, 194–195, 197, 199
petioles, 110–111
Philippines, 99
photoperiodism, 152–158
phototropism, 106
phyllotaxy, of involucel, 115–116, 217
  of leaves, 108
phylogeny, xiv, xvi, 12, 185, 208–211,
    215*
Phymosia, 186
phytoalexin, 207
phytogeography, xvi, xviii, 9, 103, 130,
    183, 202
Pijl, L. van der, 133, 135–136
pistil, 121, 123
placentation, 124
Plagianthus, 120
Pleistocene, 145, 147, 164, 200–202
pollen, 121, 123–124, 190, 199, 210
pollinators, 119–120, 124, 190–200, 221
Polynesia, 161, 171
Presl, K., 216–217
primitive characters, 135–136, 164, 172,
    185, 191–193, 208–213
Proctor, V. W., 142
productivity, agricultural, 167, 169–170,
    174–175
Prokhanov, J. I., 12–13
protein, 106, 173–175, 203
Pueblo Indians, 169
Puerto Rico, 93
Puná, Island of, 188

Quaternary, 189

Radyera, 111
Rafinesque, C. S., 5
Raven, P. H., xviii
Recent (geologic epoch), 163, 200

Redutea, 21
  heterophylla, 28
  tripartita, 30
Reeves, R. G., 4
Reichenbach, H. G. L., 3
relicts, xvii, 116, 136, 146, 158, 187, 189, 199
reproductive biology, xvii, 128, 150–157, 187, 189–200, 220
reptiles, 134, 137
resistance, to pests and diseases, 173
reticulate evolution, 209
Roberty, G., 12
Robinsonella mirandae, 107
Rohr, J. B. P. von, 5
roots, 106, 151, 185, 187–188, 190, 204
Royle, J. F., 103

Sachet, M.-H., 14, 100
Saint Croix, 5
Saint Thomas, 143
saltwater environment, 142, 147–148, 171, 187
Samoa, 70, 144
San Marcos Island, 58
Santa Isabel, 132
Sauer, J., 146, 148, 172
savanna, 114, 139, 141, 187, 189, 196
Schlechtendal, D. F. L. von, 15
Sea Island cotton, 169
sea-level fluctuation, 145, 201
seasonal cycles, 151, 155–157
seasonal flowering
  in *Cephalohibiscus*, 157
  in *Gossypium*, 151, 155
  in *Hampea*, 156
  in *Lebronnecia*, 157
  in *Thespesia*, 156
seed coat, 4, 127–129, 137, 139
  hairs, 5, 126–129, 133, 139–140, 142, 144, 147, 162, 166, 169, 171–175, 220
  impermeability of, 127–128, 139, 141–142, 144, 147, 165, 175
  outer pigment layer of, 128–129
seedlings, 106, 148
seeds, 4, 104, 105*, 125–129, 147, 172–176, 196–197, 213
  germination of, 104, 106, 127–128, 135, 141–143, 147, 164, 169, 176, 210
Seemann, B., 3

segregate genera, 15, 100, 181, 193, 202, 223
selection, and breeding, 152, 167–177, 204, 207
Selera, 39
  gossypioides, 58
self-compatibility, 148, 194, 196
self-incompatibility (self-sterility), 148, 196, 199–200
Senegal, 28, 188
senescence (of leaves), 152, 155
Senra, 198
sesquiterpenes, 203
Shantzia, 15, 84
  garckeana, 93
Shinners, L. H., 208
Short-day response, 153, 157–158. *See also* day length
Siberia, 153
sibling species, 200
similarity indices, 211, 214
Sinaloa, 56, 58
Sind, 65, 189
Skovsted, A., 16
Smith, A. C., xvii–xviii
Socorro Island, 70, 143, 146
Solander, D. C., 14
Solomon Islands, 21, 162
Somalia, 27, 65, 160, 189, 198
Sonora, 56, 58, 155
Sonoran Desert, 150, 155
South Africa, 26, 64, 132
South-West Africa. *See* Namibia
South America, 9, 29–30, 32–33, 35, 62, 71, 79, 130, 132, 142–143, 160–161, 164, 171, 188, 202
spinning jenny, 173–174
Sporne, K. R., 136
stamens, 122, 210
staminal column, 121, 122*, 123, 210
Standley, P. C., 15, 216
Stebbins, G. L., Jr., 133–134, 142, 153, 208
stem, 107–108
sterility barrier, 9, 37, 71, 161
stigma, 122*, 123–124, 192–194, 196, 199–200, 221
stipules, 101–102, 108, 110–111, 210
Sturtia, 39
  gossypioides, 48
style, 4, 124, 192, 194, 196–199

Sudan, 65
Swartz, O., 5
symbiosis, 112
sympodia, 113, 191–192
systematics, xvii, 208, 219, 220, 223–225.
    See also taxonomy

Tabasco, 75, 78
Tahiti, 144
Tamaulipas, 171
Tashkent, U.S.S.R., 12
taxonomists, xiii, 201–202, 221, 224–225
taxonomy, xiii, xv–xvi, 7–8, 12, 17, 132,
    177, 181, 225. *See also*
    systematics
Tehuantepec, 28
temperatures, night, 154
Tertiary, 153
tetraploidy, 71, 144–148, 162–165,
    169–170, 181–182, 201
Texas, 33, 142, 155
Texas A&M University, 12
Theobroma, 218
Thespesia, 14–15, 17, 19, *84* (incl. key to
    spp.), 100, 106, 123, 130–133
    (maps), 178, 180, 187, 192, 196,
    203, 211–214, 223
  sect. *Lampas*, 85, 95, 100, 132, 180
  sect. *Thespesia*, 84, *86*, 100
  acutiloba, 89, 90*, 102, 112
  altissima, 99
  banalo, 89
  beata, 93
  beatensis, 89, 101, 112, 115, 118, 126,
    140, 156, 160, 190, 196, 199
  brasiliensis, 99
  campylosiphon, 99
  cubensis, 89, 91*, 101, 107–108, 110,
    112*, 126, 133, 140, 197,
    199–200, 220
  danis, 89, 111–112, 115*, 127, 133, 156,
    179–180, 187
  debeertsii, 93
  fissicalyx, 93, 96*, 101, 107, 140
  garckeana, 93, 97*, 100–101, 109, 112,
    117, 125, 127, 142, 156, 179,
    196, 217
  grandiflora, 93, 94, 112, 118–119, 124,
    126, 128–129, 133, 140–141,
    156, 162, 190, 199, 220
  gummiflua, 93, 98*, 101, 142

  hockii, 93
  howii, 89
  lampas, 95, 101, 112, 116, 125–126,
    133, 140, 156, 162, 179, 185,
    190, 196, 217
    var. *longisepala*, 99
    var. *thespesioides*, 99
  macrophylla, 86
  mossambicensis, 89, 92*, 102
  multibracteata, 93, 95*, 116–117, 140
  patellifera, 89, 107, 124, 140
  peekelii, 20, 89
  populnea, 84, 86, 87*, 100–101, 104*,
    111–113, 115*, 117, 124,
    140–141, 147–148, 156, 162,
    179, 187, 190, 196, 203, 218
    var. *acutiloba*, 89
    var. *populneoides*, 89
  populneoides, 87, 88*, 100, 112–113,
    140, 148, 187, 196
  rehmannii, 26, 99
  robusta, 93, 96*, 107, 140
  rogersii, 93
  sublobata, 95
  thespesioides, 99, 101, 126, 139–140,
    156, 185, 187, 196
  thurberi, 56, 100
  tomentosa, 78, 100
  trilobata, 93, 109
Thespesiopsis, 15, 84, 100
  mossambicensis, 89
Thurberia, 39
  thespesioides, 55
  triloba, 55–56
timber, 162
Todaro, A., 5, 6*, 9, 13, 166, 177
toucans, 135
translocations, 182
Transvaal, 28, 166
Trifolium, 218
triploids, 182
Triticum, xvii
tropical imperative, 150–153
types, 7

Ulbrich, E., 4, 16, 162
Ulbrichia, 15, 84, 100
  beatensis, 89
Urban, I., 100
Urena, 111

Varuntsyan, J. S., 10, 12
Vavilov, N. I., 9, 12
Venezuela, 30, 143, 160
Veracruz, 75
Viability of seeds, 128, 141
Virgin Islands, 143

Waddington, C. H., 130
Wake Island, 144
Walters, S. M., 222
Watson, S., 4
Watt, Sir George, 7*, 9, 13, 166, 177

West Africa, 27–28, 172, 194
wood anatomy, 4
wood structure, of *Cephalohibiscus*, 21

Xeromorphism, 187–190
Xylon, 39

Yucatán, 28, 78, 143, 146, 171, 188

Zaitzev, G. S., 8*–13, 166, 170–171
Zaitzeva, M. G., 12
Zopilote Cañon, 190